The Great Warming

Also by Brian Fagan

Fish on Friday: Feasting, Fasting, and the Discovery of the New World

From Stonehenge to Samarkand (editor)

Chaco Canyon: Archaeologists Explore the Lives of an Ancient Society

Before California

The Long Summer: How Climate Changed Civilization

The Little Ice Age: How Climate Made History, 1300–1850

Egypt of the Pharaohs

Floods, Famines and Emperors

Into the Unknown

From Black Land to Fifth Sun

Eyewitness to Discovery (editor)

Oxford Companion to Archaeology (editor)

Time Detectives

Kingdoms of Jade, Kingdoms of Gold

Journey from Eden

Ancient North America

The Great Journey

The Adventure of Archaeology

The Aztecs

Clash of Cultures

Return to Babylon

Quest for the Past

Elusive Treasure

The Rape of the Nile

The Great Warming

Climate Change and the Rise and Fall of Civilizations

Brian Fagan

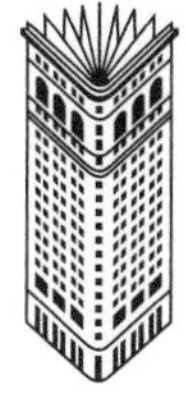

BLOOMSBURY PRESS

Published by Bloomsbury Press, New York
Distributed to the trade by Macmillan

ART CREDITS

Maps by Steve Brown; Wharram Percy medieval village reconstruction, p. 2, © by English Heritage Photo Library; *Surface Temperature Reconstructions for the Last 2,000 Years*, p. 17, reprinted with permission from the National Academies Press, © 2006 by the National Academy of Sciences; Mande Hunter and Snake Avatar, p. 84, from *The Way the Wind Blows*, ed. Roderick J. McIntosh et al., © 2000 by Columbia University Press; Double-hulled Tahitian canoe, p. 192, © by Corbis; Temperature curve for eastern China, p. 215, reprinted with permission from the National Academies Press, © 2006 by the National Academy of Sciences.

All papers used by Bloomsbury Press are natural, recyclable products made from wood grown in well-managed forests. The manufacturing processes conform to the environmental regulations of the country of origin.

LIBRARY OF CONGRESS CATALOGING-IN-PUBLICATION DATA

Fagan, Brian M.
The great warming : climate change and the rise and fall of civilizations / Brian Fagan. —1st U.S. ed.
p. cm.
Includes bibliographical references and index.
ISBN 978-1-59691-392-9 (alk. paper)
1. Global warming—History—To 1500. 2. Human beings—Effect of climate on. 3. Climatic changes—Social aspects. I. Title.
QC981. 8. G56F34 2008
904'.5—dc22 2007023092

First U.S. Edition 2008

3 5 7 9 10 8 6 4 2

Designed by Rachel Reiss

Typeset by Westchester Book Group
Printed in the United States of America by Quebecor World Fairfield

To

the Great Cat of Ra and the Venerable Bede

Black-and-whites extraordinary

"All right," said the Cheshire Cat; and this time it vanished slowly, beginning with the end of the tail, and ending with the grin, which remained some time after the rest of it had gone.

—Lewis Carroll, *Alice's Adventures in Wonderland* (1865)

Contents

Preface

"My name is Ozymandias, King of Kings,
Look on my works, ye Mighty, and despair!"
Nothing beside remains. Beyond the decay
Of that colossal wreck, boundless and bare,
The lone and level sands stretch far away.

—Percy Bysshe Shelley, "Ozymandias" (1812)

THE GREAT HOUSE, PUEBLO BONITO, stands gaunt and silent, nestling under the precipitous cliff, the serried rooms open to the gray sky. A chill wind scatters dead leaves and delicate snowflakes across the empty plaza on this bleak winter day. Clouds hang low over the cliffs of Chaco Canyon, New Mexico, swirling in the gusts of the January storm. The silence is complete.

A thousand years ago, Pueblo Bonito was a sacred place, which echoed to spectacular dances at the summer solstice. Visitors from miles around flocked to this, perhaps the greatest of all southwestern pueblos. Then, in A.D. 1130, fifty years of drought sank over Chaco Canyon. Maize yields plummeted. Within a few years, Pueblo Bonito emptied. Half a century later, Chaco Canyon was virtually deserted. After many centuries within the canyon walls, the Ancestral Pueblo had moved away and settled with relatives living in better-watered areas.

This winter day, no ghosts of a thousand years ago rise to haunt my imagination and excite my consciousness. The past is dead, long vanished into oblivion. I'm reminded of Shelley's Ozymandias, King of Kings, his deeds forgotten, his palaces reduced to crumbling ruins.

In A.D. 1118, a decade before the great drought arrived at Chaco, the Khmer god-king Suryavarman II ascended to the throne of Angkor on Cambodia's Tonle Sap in Southeast Asia. Almost immediately, he began building his masterpiece, Angkor Wat. Thousands of his subjects labored on his palace and temple, a vast replica of the Hindu universe, complete with sacred mountains. Nothing mattered but to serve the god-king. Suryavarman and his successors created a centripetal religious utopia erected on a foundation of intensive rice cultivation, irrigated with canals, reservoirs, and flooded paddies nourished by the summer flood.

Angkor Wat no longer boasts gilded towers and brilliantly painted temples. But it still mesmerizes, with its maze of stairways and long, echoing galleries adorned with yard upon yard of royal processions, armies on the march, and sinuous dancing girls promising the delights of paradise. Then you realize that the place is lifeless, a moment frozen in time, abandoned by its builders when in full magnificence, partly because drought dried up their rice paddies and they went hungry.

Again, Ozymandias comes to mind. Angkor Wat leaves you with a sense of futility and despair.

Chaco Canyon and Angkor Wat are silent testimony to the power of climate to affect human society, for better or worse.

Soon after Suryavarman's loyal subjects labored on Angkor Wat, the cathedral of Notre Dame de Chartres rose in northern France. Built in a mere sixty-six years after A.D. 1195, this Gothic cathedral was the sixth church on the site, a miracle in stone and glass. Like Angkor Wat, Chartres is a masterpiece, but this one's still part of the fabric of human life, a place where masses are celebrated and psalms chanted. Here the infinite becomes a miracle in stone and glass. Chartres is all windows, set among soaring beams and graceful arches. Gemlike sunlight shines through them, creating transcendental effects. The setting still brings heaven to earth and links the secular and the spiritual, just as it did a thousand years ago. Here the past is still alive.

Chartres was built at a time when Europe basked in a warmer climate and enjoyed a long series of good harvests. Those who benefited thanked God and the unknown powers of the cosmos for their bounty. They built a cathedral in gratitude.

The world of a thousand years ago was a vibrant, diverse place, much of it a tapestry of volatile civilizations, great lords, and endemic warfare. Camel caravans, the Great Silk Road, and monsoon winds connected much of the Old World in the first iteration of a truly global economy. However, most humans still lived in small hunting bands or as subsistence farmers, surviving from harvest to harvest, eking out a living from the soil. We have long known of this world from archaeology, from excavations into great cities, into caves and humble shell mounds, and from scatters of Norse iron nails in the High Arctic, from historical documents and oral traditions. But it's only now that we're learning just how profoundly the warmer climate of the day affected humanity.[1] This book is the story of five centuries of changing climate—in fact, of a global warming—between A.D. 800 and 1300, and of the changes' impact on the world of a millennium ago. As in our own time, climate change did not plot a straight line from year to year, and varied from place to place. But its peaks and valleys followed a trend that we can clearly make out in retrospect. We have much to learn from this story about the power of climate change to affect our own future.

THE MEDIEVAL WARM Period was named half a century ago by a British meteorologist, Hubert Lamb.[2] He wrote of an era from about A.D. 800 to 1200 that he pieced together from a jigsaw of climatological and historical clues: four or five centuries of relatively amiable climate that brought good harvests to Europe and permitted the Norse to land in Greenland and North America. The Medieval Warm Period gave way to six centuries of highly unsettled climate and cooler conditions: the Little Ice Age.

We have long known many details of the better-documented Little Ice Age, when, famously, the Thames River froze over.[3] There were famines and severe tempests, occasional winters of exceptional cold. But the Medieval Warm Period was, until recently, a climatological mystery. Lamb wrote at a time when paleoclimatology, the study of ancient climate, was in its infancy, and long before humanly caused global warming was on the scientific radar. Today, we know a lot more about the Medieval Warm Period than he did. Thanks to tree-ring research, we now have detailed information on seasonal rainfall and temperatures in Europe and in the North American Southwest going back at least a millennium. In the chapters that follow, sidebars will discuss some of the methods we use to study prehistoric climate. Ice cores from Greenland, also from high in the Andes and elsewhere, provide important data on cooler and warmer cycles over the past two thousand years. Growth layers in tropical coral from small Pacific atolls also document climatic shifts over many centuries. And tree-ring sequences from around the world are slowly putting flesh on the Medieval Warm Period's still-shadowy skeleton.

Europeans built cathedrals and the Norse sailed to North America during the Medieval Warm Period, but the picture of the warm centuries that's emerging from the new research depicts a climatic villain as much as a hero. There was indeed warming, in most places reflected in milder winters and longer summers, but the temperature differences never amounted to more than a few degrees. Nor was everywhere necessarily warmer. In the eastern Pacific, the same centuries were cool and dry. These were times of sudden, unpredictable climatic swings, and, above all, of drought. Extended medieval dry cycles helped topple Chaco Canyon and Angkor Wat, contributed to the partial collapse of Maya civilization, and starved tens of thousands of northern Chinese farmers.

Much of this aridity can be attributed to persistent La Niña conditions in the Pacific, especially around 1100 to 1200, but climate change was not the only villain. (See the sidebar in chapter 9 for a discussion of La Niñas.) No one in their right mind would argue that climate "caused" all the economic, political, and social changes described in these pages. That kind of environmental determinism, the notion that climate caused the major developments of history, was discredited more than three quarters of a century ago. The effects of climatic shifts were usually far more indirect.

While I was writing this preface, I went for a walk along the shores of a nearby slough. I picked up a small pebble and threw it into the mirror-still water. A plop and the stone disappeared, but the concentric ripples from the point of impact radiated outward toward the banks. A surprisingly long time passed before the last ripple vanished. So it was with ancient climate change. It was not so much the immediate impact of a major shift like drought, a flood cycle, or an El Niño that caused political or social change. Rather, the subtle consequences that rippled through society made the difference: new strategies for storing water; the planting of more drought-resistant cereals; the development of new institutions such as secret societies that collected information for predicting rainfall. This book is as much about how the human societies of a thousand years ago coped with climate change as it is about the warming and other climatic phenomena themselves.

Humans have always lived in unpredictable environments, in a state of flux that requires them to adapt constantly and opportunistically to short- and long-term climate change. What is fascinating about the world of a thousand years ago is that we now have just enough climatological information to look behind the scenes, as it were, to examine the inconspicuous undercurrents of climate that helped push Angkor to collapse or forced Mongolian horse nomads to search for new pasture. These undercurrents are now part of the meat and drink of history. A generation ago, they would have been ignored.

The Great Warming explores societies both historically obscure and well known. We cannot understand the significance of the Medieval Warm Period without traveling far beyond Europe, where the effects of the warmer centuries were strongly positive and the continent saw the cultural flowering we now call the High Middle Ages. The higher temperatures and accompanying shifts in rainfall patterns rippled across the globe, bringing both opportunity and catastrophe.

One consequence was greater interconnectedness between radically different societies living long distances apart. The Norse took advantage of

Some Major Historical Events

This is a selective listing in chronological order. For clarity, many major events and developments are omitted.

A.D. 570	Birth of Mohammed
600	Classic Maya civilization in full swing
618	T'ang dynasty begins in northern China
710	Islamic conquest of Spain (al-Andalus)
750	Abbasids assume power in Baghdad, ushering in a great era of Islamic rule and scholarship
793	Norse raid on Lindisfarne, England
802	Jayavarman II founds the Angkor state, Cambodia
814	Death of King Charlemagne of France (742–814)
874	Norse settlement of Iceland
900	Sicán lords dominate Peru's north coast; Collapse of Maya civilization in southern lowlands
907	T'ang dynasty falls in northern China; Period of Khitan conquests begins in Manchuria and Mongolia
971	Mahmud, a Ghazi ruler from Afghanistan, raids India for sixty years
980s	Eirik the Red colonizes Greenland
990s	Norse discovery of North America; sporadic trade with Inuit groups in the Baffinland area
1000	Thule population movements from the Bering Strait eastward across the High Arctic and to Greenland
1066	William the Conqueror invades England: the Norman Conquest
1100	Gradual abandonment of Chaco Canyon, New Mexico
1113	Suryavarman II starts building Angkor Wat, Cambodia

1181	Jayavarman VII builds Angkor Thom, Cambia
1200	Chimor controls Peru's North Coast; First settlement of Rapa Nui (Easter Island); First settlement of New Zealand occurred at an unknown date around this time
1206	Ginghis Khan elected Great Khan of the Mongols; Muslim dynasties assume power in Delhi, India
1207	Mongolian campaigns against the Chin of northern China begin
1215	Ginghis Khan captures Beijing
1220	Ginghis Khan smashes the Khwarezmed empire
1227	Death of Ginghis Khan
1230	Major El Niño event devastates Peru's north coast
1241	Mongol general Subutai defeats Henry the Bearded at Legnica, Silesia, then retreats to the steppes
1258	Mongols capture Baghdad
1276	Great drought in American Southwest lasts a quarter century; Mesa Verde is abandoned
1279	Kubilai Khan becomes emperor of China and rules until 1294
1315	Seven-year famine begins in western Europe
1324	Mansa Musa of Mali visits Cairo on his pilgrimage to Mecca
1348–	The Black Death devastates medieval Europe
1398	Timur attacks and overruns Delhi, India
1431	Angkor state collapses
1470	Chimor falls to the Inca
1492	Spanish monarchy conquers Islamic Spain; Christopher Columbus sails to the Indies
1519	Hernán Cortés lands in the Aztec empire
1532	Francisco Pizarro advances on the Inca

favorable ice conditions in the north to journey to Iceland, Greenland, and beyond, where they came in contact with Inuit hunter-gatherers in Baffinland. El Niño activity in the Pacific occasionally reduced the strength of prevailing northeasterly trade winds. Polynesian sailors sailed north and eastward to colonize some of the remotest islands on earth. During the warm centuries, more and more of Europe's gold crossed the Sahara on camels from West Africa. Strong southwesterly monsoons fostered nonstop voyages across the Indian Ocean from the Red Sea, Arabia, and East Africa to India and beyond. All these long-distance connections changed history, as did many other interconnections that ebbed and flowed with the changing political fortunes of human societies and climatic shifts.

Opportunity went hand in hand with misfortune. When we move beyond Europe and the North Atlantic into drier environments and lands with unpredictable rainfall, we enter a medieval world where drought cycles and even a few inches of rain could make all the difference between life and death. While Europe basked in summer warmth and the Norse sailed far west, much of humanity suffered through heat and prolonged droughts. A huge swath of the world, from much of North America through Central and South America, and far across the Pacific to northern China, experienced long periods of severe aridity. Drought cycles settled over the Saharan Sahel, the Nile Valley, and eastern Africa, creating havoc. Farmers went hungry, civilizations collapsed, and cities imploded. Archaeology and climatology tell us that drought was the silent killer of the Medieval Warm Period, a harsh reality that challenged human ingenuity to the limit.

Most societies on earth were affected by medieval warming, many of them for the worse.

TODAY, PERHAPS EVEN more than a millennium ago, we live in climatically dramatic times: we have seen a steady rise in global temperatures accompanied by weather-related disasters from tsunamis to hurricanes. While the scientists quietly labor in the background, the chatterers and doomsday sayers are vociferous in their predictions of disaster triggered by anthropogenic global warming. Yet almost none of these self-proclaimed prophets bother to look back at climate change in earlier centuries and millennia, except for politically charged discussions as to whether the world was warmer a thousand years ago than it is today. It wasn't; we've entered a time of sustained warming, which dates back to at least 1860, propelled in large part by human activity—by the greenhouse gases from fossil fuels.

The prolonged debate over anthropogenic global warming is over, for the scientific evidence documenting our contributions to a much warmer world of the future is now beyond the stage of controversy. Now the discussions are changing focus, as we grapple with the long-term problems of reducing pollutants and living with the consequences of a world where ice sheets are melting and sea levels rising. Most of the passionate debate about contemporary climate change revolves around extreme weather events and sea level rises. The melting of ice caps and the increased danger of flooding are no trivial matter. But the experience of the Medieval Warm Period tells us that the silent and oft-ignored killer is drought, even during a period of mild warming. The computer projections for drought in an anthropogenically warmed world, described in chapter 13, are frightening. We already know that some 20 million to 30 million tropical farmers perished as a result of droughts during the nineteenth century, when there were far fewer people on earth.[4] Now we are entering a period of sustained warming with millions of people already at risk, living as they do on agriculturally marginal lands, or, in the case of Arizona and California, in huge cities looting water from aquifers and rivers.

The Medieval Warm Period tells us much about how humans adapt to climatic crisis, and offers forewarning of lengthy droughts when warming occurs. We're entering an era when extreme aridity will affect a large portion of the world's now much higher population, where the challenges of adapting to water shortages and crop failures are infinitely more complex. One can only hope that our uniquely human qualities of adaptiveness, ingenuity, and opportunism will carry us through an uncertain and challenging future.

Author's Note

Place names are spelled according to the most common usage.

The names of archaeological sites and historic places are spelled as they appear most commonly in the sources I used to write this book. Some obscure locations are omitted from the maps for clarity; interested readers should consult the specialist literature.

The notes tend to emphasize sources with extensive bibliographies, to allow the reader to enter the more specialized literature if desired. This being a historical narrative, sidebars at intervals in the text provide further information on such phenomena as the Intertropical Convergence Zone and major climatological methods.

All radiocarbon dates have been calibrated and the A.D./B.C. convention is used.

Temperature curves have been smoothed statistically for clarity.

CHAPTER 1

A Time of Warming

The occurrence in medieval York of the bug Heterogaster urticae, *whose typical habitat today is stinging nettles in sunny locations in the south of England, discovered by . . . archaeological investigation to have been present there in the Middle Ages . . . presumably indicates higher temperatures than today's.*

—Hubert Lamb, *Climate History and the Modern World* (1982)[1]

SOUTHERN ENGLAND, FALL, A.D. 1200. The chilly mist hangs low over the treetops. A pervasive drizzle drifts across the plowed strips, misting the weathered faces of the two men sowing wheat from canvas seed bags slung around their necks. Snub-nosed and tousle-haired, they are barefoot, clothed in dirty, belted tunics and straw hats, swinging effortlessly to and fro, casting seed across the shallow furrows. Behind them, an ox-drawn harrow, a square wooden frame with wooden spikes pointing into the earth, covers the newly planted seed. As one strip is sown, the men move on to the next, for time is short. They must plant before heavy autumn rains can wash the seed from the earth.

The routine of planting, learned in childhood, is as unchanging as the passage of the seasons. Older men remember cold, dreary days

English medieval farmers sow grain, then harrow it into the ground (top)*. Women reap and bundle the grain at harvest* (bottom)*. (Reconstructions based on excavations at Wharram Percy, northeastern England.*

when even a sheepskin cloak could not keep out the pervasive chill. They also recall years when the sun blazed down from a cloudless sky, the heat shimmering above the fields. These were times when the village gambled that it would rain and planted anyhow. Sometimes, the bet paid off. All too often, it did not. When it didn't, there would be hunger the following year.

Their seed bags empty, the two men stretch and sling new ones from their shoulders. They're tired after days of backbreaking labor, harvesting summer crops, then plowing and planting for winter wheat. The work never ceases in a farming world where everyone lives on the edge, where the unspoken threat of hunger is ever present.

The village is coming off a good summer harvest after weeks of fine weather, so there is plenty to eat. Good fortune continues. The winter is mild and not too wet. January and February bring frost, even a little snow; but there are no late cold snaps, and spring is early, with warm temperatures and just enough gentle rain. As the days lengthen, the villagers weed the growing crop. By late July, the grain is ripe and the harvest begins. The fields bake in the hot sun, a deep blue sky with fluffy clouds overhead. The men bend to the harvest. They reap the ripe wheat with short iron sickles, gathering bundles of stalks in their hands and cutting them, only pausing to sharpen their blades. Behind them, the women have tucked their skirts into their belts to free their legs. Brightly colored cloths cover their hair. They bind and stack the sheaves in the field, but soon the grain will be carried indoors, stored on the stalk for threshing and winnowing under cover when the weather turns bad. Children play among the sheaves and gather grain among the stubble. The workers pause at midday to stretch their stiff backs, to drink some ale, as birds jostle and swoop overhead. Soon the harvest will resume, continuing until dark as the village races against time to gather in the crop while it is dry.

To judge by today's subsistence farmers, those of A.D. 1200 would let nothing go to waste, even in a good harvest year like this one. You have only to glance at the deep wrinkles etched into adult faces to get the point. Even men and women in their twenties look old, their countenances withered by brutal hard work and occasional hunger or malnutrition. Yet these people lived in a world that was warmer than it had been for some centuries, in what climatologists call the Medieval Warm Period.

A THOUSAND YEARS ago, everything in Europe depended on agriculture. From Britain and Ireland to central Europe, 80 to 90 percent of the population struggled to coax a living—and, if they were lucky, a food surplus—from the soil. Europe was a continent of subsistence farmers, who lived from harvest to harvest, whose fate depended on the vagaries of temperature and rainfall.

There were many fewer people then. The population of London exceeded 30,000 for the first time in 1170, a vast metropolis by the standards of the day. Other English population centers were much smaller.

Norwich in East Anglia, for example, had but 7,000 to 10,000 inhabitants. The combined population of France, Germany, Switzerland, Austria, and the Low Countries was about 36 million in 1200, compared with over 250 million today. Nearly all these people lived in hamlets and villages, or perhaps small towns, for cities were only just becoming a significant element in European life. And everyone, even the greatest lord, depended on a countryside farmed without machines, hybrid seed, or fertilizers. Horses and oxen, even wives, hauled the plow and the harrow. The harvest was gathered by hand, carried on people's backs, perhaps transported by oxcart or river barge to market.

The rural landscape was a mosaic of forest and woodland, river valley and wetland, modified constantly by human activity. Many people lived in small, dispersed settlements, surrounded by haphazard fields. But increasingly, they dwelled in larger, more centralized villages, where the nearby arable land was divided into large, open fields in turn subdivided into small strips of about a half acre (0.2 hectare) each. Each tenant had possession of several groups of strips, often called furlongs, but not all of that land was under crops at once. Every farmer knew that arable land had to be grazed and manured by animals, then rested to regain its fertility and minimize plant diseases. The best-drained, lightest soils supported cereal crops. Animals grazed not only on stubble, but in the woods and in open pasture on heavier, more clayey soils. Like modern-day subsistence farmers in Africa, medieval European peasants knew the properties of different grazing grasses, the subtle indicators of renewed soil fertility, the seasons of wild plant foods. Their only protection against sudden frosts, storms, or drought was a diverse food supply based on far more than cereal crops.

Coaxing a living from Europe's medieval soils was never easy, but it was done, and sometimes with considerable success, especially during runs of warm, drier summers. Farmers in England and France grew mainly wheat, barley, and oats. As a gross generalization, about a third of the land was planted in wheat, half in barley, the rest in other crops, including peas. Even in good years, the yields were small by today's standards. A good wheat harvest would yield between 8 and 12.5 bushels (2.8

to 4.0 hectoliters) per acre (0.4 hectare). Compare this with today's figure of over 47 bushels (16.5 hectoliters) an acre. When you realize that 2.3 bushels (0.8 hectoliters) of that yield went back into the soil as seed for the next crop, the yield was small indeed, leaving very little chance of a food surplus in any but the best years. The figures for barley, used for beer, were somewhat higher (23.5 bushels/8.3 hectoliters), but the amount of seed planted was greater. In good years, grain yields of slightly under four times seed grain sown were the norm. One survived by diversification.[2]

Everyone grew vegetables. Protein-rich peas and beans were planted as field crops in early spring and harvested in fall; the legumes were allowed to dry on the plant, the stalks plowed back into the soil as fertilizer. Vegetables and herbs of all kinds supplemented what was basically a meatless diet based on bread and gruel.

Most farmers had a few head of livestock—a milk cow or two, some pigs, sheep, goats, and chickens, and, if they were lucky, a horse or some oxen, or at least access to them for plowing. Animals provided meat and milk, also hides and wool. Sheep shearing was an important occasion in spring, undertaken on a carefully chosen fine day when a warm wind from the west brought promises of summer. The breeze sends wood smoke cascading out of windows and doors, opened by the women to let in fresh air. Outside, the village flock crowds in a large wicker pen, sheep jostling against one another. A smell of wool fills the air. The men, dressed in leather jerkins, grab the sheep one by one and shear them with simple iron clippers, deftly flipping the docile beasts on their backs to complete the task. The shorn, bewildered beasts shake themselves as young boys drive them to a nearby corral. Hovering children gather up the wool and lay it on wooden racks to dry in the bright sunlight.

Most of the year, animals grazed and foraged on their own—this was especially true of pigs, which feasted off acorns and beechnuts in the fall. But winter feeding was another matter, the challenge being to keep breeding stock alive. Surplus males and dried-up cows no longer yielding milk were either sold or slaughtered in the autumn to free up hay for the most valuable animals. The hay harvest was all-important. Mowing began in June and continued into July, depending on the weather, for the hay had to be absolutely dry, lest it rot after harvest and become so hot that it would catch fire. On fine days, men with long-bladed iron scythes worked their way across a meadow, leaving the crop to dry in rows across the field. They would return and turn the crop over a couple of times to dry better before stacking it in such a way that the outer layer formed a thatch to keep off the rain. The hay harvest was a major event in the year but so dependent on dry conditions that a wet year could lead to stock losses the next winter, perhaps the loss of every beast. Once again, everything depended on the climate.

Even in a bad year, the farmer still had to pay taxes and church tithes, which ate into food supplies. A man with a wife and two children could survive at a basic level with 5 acres (2 hectares). But everyone, even young children, had to grow vegetables and forage for wild foods such as mushrooms, nuts, and berries. Five acres left precious little margin for poor harvests caused by frost or storms. A succession of years meant famine, famine-related diseases, at best malnutrition, and certainly some dying, especially in the cold and miserable months of late winter, when food supplies were always low and Lent with its fasting was just over the horizon.

Each year, as summer ripened into fall, every community reaped its harvest and gave thanks to God for his bounty, for life was never easy. The endless cycle of the seasons defined human existence. So did the routines of planting, growth, and harvest; the verities of birth, life, and death; and what everyone believed were the arbitrary whims of the Lord.

In an era long before long-range weather forecasting, everyone, whether king or noble, warlord, merchant, or farmer, was at the mercy of cycles of heavier rainfall and drought, savage gales and perfect summer days. They were unwitting partners in an intricate climatic gavotte between the atmosphere and the oceans. But, especially between A.D. 800 and 1300, the dance slowed slightly into a measured waltz, where summer warmth and more settled conditions tended—and one stresses "tended"—to be the norm. The gyrations of climate change slowed momentarily. Europe changed profoundly during these five centuries from 800 to 1300, the Medieval Warm Period.

IN THE LARGER scheme of things, the twenty generations or so of medieval warming are but the blink of an eye. The relatively minor temperature changes of these centuries pale alongside those that occurred at the end of the most recent ice age. About twelve thousand years ago, the world entered a period of sustained global warmth, known to geologists as the Holocene (after the Greek words *holos*, "whole," and *kainos*, "new," the word thus meaning "entirely recent"), which continues to this day. Generations of scientists, working with inadequate data, conjured up images of over ten millennia of a basically modern climate that had changed relatively little since the warm-up after the Ice Age. But a revolution in paleoclimatology, the study of ancient climate, has transformed our knowledge of the Holocene in recent years.

Today's climatologists drill into sea and lake beds, take deep cores from Greenland and Antarctic ice sheets, and pore over tree-ring sequences taken from the trunks of ancient trees. Their research has revealed a Holocene climate constantly on the move. We can now discern not only millennium-long cold and warm oscillations, but also much shorter cycles, especially over the past two thousand years. The shifts from slightly wetter to slightly drier, from warmer to cooler and back again, never cease. Some endure a century or a decade; others, like El Niño events, last no more than a year or so. Few major climatic events lay within the span of generational memory, and were thus quickly forgotten in times when life expectancy everywhere was little more than thirty years. The new climatology has shown us that the climatic timepiece may accelerate and slow down, falter and change direction suddenly, even remain steady for long periods of time, but it never stops.

No one knows exactly what drives the climatic pendulum. Most likely, small changes in the earth's obliquity trigger climate changes. So do cycles of sunspot activity. For instance, a dearth of sunspots during

Studying Ancient Climate Change

Archaeologists, historians, and paleoclimatologists use a wide variety of methods to study ancient climate change. Here are the principal ones:

Direct Methods

INSTRUMENT RECORDS

Instrument records are the most accurate and direct way of studying climate change. Unfortunately, such archives go back only some 150 years in Europe and North America and for much shorter periods elsewhere.

HISTORICAL DOCUMENTS

Archives provide invaluable snapshots of ancient climate, from such documents as diaries, logs, and official reports that mention contemporary events such as floods or droughts. The oldest are the reports of the flowering of cherry trees in Japan and Korea, which date back a thousand years. In Europe and the Mediterranean region, records for many areas go back to about 1500.

Indirect Methods (Proxies)

ICE CORES

Deep cores drilled into ice sheets such as those in Greenland, Antarctica, the Andes, and Tibet provide continuous records of temperature changes derived from measurements of oxygen- and hydrogen-stable isotopic ratios in the water molecules that make up the ice. Such changes in ratios can be connected to temperature shifts. One ice core from Antarctica takes the record back over 420,000 years. High-resolution sequences for the past two thousand years come from Greenland, the Andes, and elsewhere.

DEEP-SEA AND LAKE CORES

Marine sediments recovered from deep-sea cores contain temperature-sensitive foraminifera and marine diatoms and can go back tens of thousands of years. In some locations, like the Cariaco basin off Venezuela, and California's Santa Barbara Channel, rapid accumulation rates have provided relatively precise records of medieval warming and of later cooling. Lake cores yield seasonal layers that record changes in water balance, and hence information on ancient droughts.

CORAL RECORDS

Corals living near the sea surface produce annual density bands of calcium carbonate. By measuring the changing ratios of O-18 to O-16, researchers can detect temperature changes, the ratio decreasing with increasing warmth. Coral records tend to be incomplete. Few go back more than two or three centuries.

TREE RINGS (DENDROCHRONOLOGY)

Dendrochronology is based on the study of the annual growth rings of trees, the thickness of the rings providing a proxy for rainfall shifts. Originally developed in the American Southwest, tree-ring researches now provide important proxy data in many parts of the world. Records from Europe are notably comprehensive, as are those from parts of North America. In recent years, efforts have been made to collect more samples from Asia and the southern hemisphere, which promise to throw important light both on the Medieval Warm Period and on ancient El Niño events. Tree-ring records go back almost to the Ice Age in Europe, but are generally most common for the past one thousand to two thousand years.

These are the major paleoclimatological proxies. Others include cave deposits such as stalagmites, which record the changing isotopic composition of cave groundwater and temperature through time, and temperature information derived from boreholes.

Climatic Forcings

Forcings are powerful and unusual factors like volcanic eruptions that can cause climate change. In the context of the Medieval Warm Period, these are natural changes such as solar irradiance caused by small tilts in the earth's orbit and by major volcanic eruptions that affect global energy balance. Large volcanic events add large amounts of ash and sulfur gases to the atmosphere, diminishing the amount of solar radiation reaching the earth and thereby cooling it. The effects are limited to a few years. Since 1860, the major climatic forcing has been humanly caused, in large part by the use of fossil fuels.

Computer Modeling

Sophisticated computer models simulate the behavior of the world's climatic system, using increasingly large quantities of raw data derived from buoys, instrument records, proxies, and satellites. Computer models are used both to understand the natural variability of global climate and to measure the effects of different forcings. They provide the basis for assessing the effects of anthropogenic global warming and both short- and long-term weather forecasts.

the seventeenth century marked a period of markedly cooler climate during the height of the so-called Little Ice Age. There are other climatic forcing agents, too, among them volcanic activity in Iceland, Southeast Asia, and elsewhere. A massive eruption of Mount Tambora on Java's Sumbawa Island in 1815 blew off 4,250 feet (1,300 meters) of the summit of the volcano. Huge clouds of volcanic ash rose into the atmosphere, masking the sun and causing Europe's celebrated "year without a summer" in 1816.[3] In recent years, however, most climatologists have come to believe that complex, yet still little understood, interactions between the atmosphere and the ocean play a major role in climatic shifts. The climatologist George Philander calls it a dance between very different partners, one fast-moving, and the other clumsier. He writes: "Whereas the atmosphere is quick and agile and responds nimbly to hints from the ocean, the ocean is ponderous and cumbersome."[4] We dance along with these partners, opportunistically, sometimes decisively, and very often with reluctance.

We've learned, too, that the gyrations of the climatic dance have a startlingly direct effect on human societies, like the exceptionally heavy rains caused by a massive El Niño that destroyed generations' worth of irrigation canals in riverbeds along Peru's north coast in the sixth century A.D., or the major drought cycles in the American Southwest that triggered population movements by Ancestral Pueblo families over a wide area a thousand years ago. At the same time as the Pueblo of the Southwest were moving away from their homes in the face of drought, Europe's medieval farmers were basking in more predictable weather conditions and ample, but usually not excessive, rainfall. The effects of the slightly warmer and drier conditions appeared in all kinds of subtle ways—in better harvests, in population growth and accelerating deforestation, in an explosion of trade and deep-sea fishing, and in a veritable orgy of cathedral building. This is not to say, of course, that greater warmth *caused* all these changes; far from it. What's exciting is that we can now begin to link seemingly minor climatic shifts to all kinds of historical events in ways that were unimaginable even a generation ago. With a few notable exceptions, like the Swiss historian Karl Pfister, who has spent years studying the dates of wine harvests, most historians have tended to ignore climatic shifts, largely because as nonscientists they were unversed in the new climatological data. Today, we can see that climate change was one of many significant factors involved in shaping medieval history, especially the lives of ordinary people living in small villages, growing crops, or fishing in the North Sea.

In about 1120, the monk and historian William of Malmesbury traveled through the Vale of Gloucester in England's West Country and admired the fertile summer landscape. "Here you may behold highways and publick roads full of Fruit-trees, not planted, but growing naturally," he wrote. "No county in England has so many or so good vineyards as this, either for fertility or for sweetness of grape. The wine has no unpleasant tartness or eagerness; and is little inferior to the French in sweetness."[5] William observed that the grapes were planted in the open, trained up on poles and not protected against cold winds by strategically placed walls. At the time, climatic conditions were ideal. Vines need freedom from spring frosts, especially at, or after, flowering, as well as sufficient sunshine and warmth in summer, not too much rain, and enough autumn sunshine and warmth to raise the grapes' sugar content. Numerous vineyards flourished in England at the time, considerably farther north than the northernmost vineyards in France and Germany during the 1960s. During the twelfth and thirteenth centuries, England's climate was so temperate that her merchants exported large quantities of wine to France much to the consternation of French growers, who complained loudly. Not that England was alone in its winemaking. Between 1128 and 1437, wine was produced in eastern Prussia at 55 degrees north, also in southern Norway. The Black Forest had vineyards up to 2,560 feet (780 meters) above sea level. Today, the highest vineyards in Germany are at 1,800 feet (560 meters). At the time, summer temperatures were 1.8 to 2.6 degrees F (1.0 to 1.4 degrees C) above those of half a century ago in central Europe, fractionally lower in England.[6]

We first learned about these warm centuries from the work of the British meteorologist and climatic historian Hubert Lamb, one of the little-known heroes of climatology. He studied the minutiae of climate change over the past two thousand years during the 1950s and 1960s, at

Locations mentioned in chapters 1 and 2. Some minor places are omitted for clarity.

a time when most historians denied that temperature and rainfall played any part in shaping historical events. Lamb was a brilliant weather detective, who had few modern-day proxy records like tree rings or ice cores to work with. Instead, he relied on scattered geological clues and on a broad array of historical records, which he fitted into a complex jigsaw puzzle, while working back in time from two hundred years or more of instrument observations throughout Europe. Among his astonishing achievements were detailed accounts of major storms in the English Channel and North Sea. For instance, Lamb reconstructed four fierce storm surges in about 1200, 1200–19, 1287, and 1382 that killed at least one hundred thousand people along the Dutch and German coasts. His recounting of the massive Atlantic depressions that overwhelmed the Spanish Armada in 1588 is a masterpiece of climatological detective work.[7] Climatologists still cite Lamb's work with respect, as they do that of the French historian Emmanuel Le Roy Ladurie, who wrote one of the first climatically driven accounts of European history, based in large part on the dates of wine harvests over several centuries—early in warm years, much later in cool, wet ones.[8]

Much of Lamb's early work was well-reasoned extrapolation. For instance, he used fifty-year averages of summer wetness and winter mildness indices, derived from records dating back to as early as 1432, to reconstruct the climate in medieval times and even earlier. He identified four centuries of significantly warmer climate after 800 that he named the Medieval Warm Period (sometimes called the Medieval Climatic Anomaly). He never thought of this as a block of time when Europe basked in warm sunlight. Rather, it was a period of cyclical fluctuations with occasional very cold winters like that of 1010–11, which gripped even the eastern Mediterranean in intense cold.

Few winters were so frigid over the next three centuries. However, the persistent warmer conditions melted ice caps, raised mountain tree lines, and caused significant sea level rises of 24 to 31 inches (60 to 80 centimeters) in the North Sea, sufficient to cause catastrophic flooding when high tides coincided with storm surges.[9]

Even without storms, the sea level rise altered the configuration of low-lying coasts. For instance, the Fenland of eastern England is a once glacial landscape of marshes, swamps, and turgid streams. This was remote, inaccessible country. As recently as the early twentieth century, the Fens supported eel fishers and marsh dwellers who lived in a world apart from the farmers that surrounded them. The Fenland was both a rich food source and a strategic advantage for those who knew how to use it. The Saxon chieftain Hereward the Wake held out at the abbey of Ely, in the heart of the Fens, against William the Conqueror for five years after the Norman Conquest of 1066. He and his men hid among a maze of obscure willow-girt islands. When William captured Ely in 1071, Hereward simply melted away into hiding, to vanish from history.[10]

The North Sea continued to rise after 1000. In Great Britain, a tidal inlet extended as far inland as Norwich. In William the Conqueror's time, the town of Beccles, now far from the North Sea, was a thriving herring port. Before the Conquest, local fishers had supplied thirty thousand herring annually to the nearby abbey of St. Edmund. William doubled the assessment. Great storms in 1251 and 1287 inundated huge tracts of the Netherlands to form a huge inland water, the Zuider Zee, as thousands of acres of coastal Denmark and Germany also vanished under the ocean.[11]

Hubert Lamb dated the culmination of warmth over wide areas to different times. There was significant warming in Greenland from about 900 to 1200. Europe experienced its warmest temperatures between 1100 and 1300, when dry summers and mild winters were the norm.

As happens with ideas in academic circles, the Medieval Warm Period became a fixture in the scholarly literature: five centuries when Europe basked in idyllic summers. Medieval warming became a dim background noise to the larger events of history, but few historians investigated the phenomenon more closely while climatic proxies were still in their infancy. Hubert Lamb himself never thought of the Medieval Warm Period as a finite segment of time, for he was well aware of the realities of European climate. A half century of climatic detective work since Lamb's research has shown that he was right. There was warming, especially between 1100 and 1200, but the climate was, as ever, infinitely variable. The Medieval Warm Period was not a discrete episode when climate was distinctly different from what came before; nor, indeed, was the Little Ice Age that followed it from about 1300 (the beginning date is uncertain) to 1860. Nonetheless, as William of Malmesbury's praise of English wine attests, an average rise of even a degree or two can change the face of a landscape or dash a civilization.

Reconstructing the climate of the Medieval Warm Period has assumed pressing importance in many scholars' minds as part of the politically charged and once controversial debate over the reality of humanly caused global warming. Those who oppose the concept of anthropogenic global warming compare the temperature curves for the warm centuries to the persistent, almost straight-line warming since the height of the Industrial Revolution in 1860.

The controversy erupted when three climatologists, Michael Mann, Raymond Bradley, and Malcolm Hughes, published a reconstruction of northern hemisphere temperatures for the past six hundred years, then for the past thousand years, using proxies such as tree rings, ice cores, and coral, as well as instrument records from the past 150 years.[12] Their seesawlike graph with its stark record of rising temperatures since 1860 received widespread attention when it was published in the report of the Intergovernmental Panel on Climate Change in 2001.[13] The Mann curve is popularly known as the Hockey Stick, on account of the long, almost straight-line warming it shows over the past 150 years. Compared with present-day warming, the temperatures of earlier centuries are almost flat, which is what infuriated those who deny humanly caused global warming. They would have one believe that the Medieval Warm Period was warmer than the present-day climate.

How warm were these five centuries in the northern hemisphere and were they, in fact, hotter than today? We know from instrument data since 1861 that winter temperatures have warmed by about 1.4 degrees F (0.8 degrees C) and summer ones by about 0.8 degrees F (0.4 degrees C). For earlier centuries, we have to rely on proxies and the occasional historical record, like those used by Hubert Lamb. Numerous proxies take us back to 1600 and show that the seventeenth century was cool, with summer temperatures about 0.9 degrees F (0.5 degrees C) cooler than between 1961 and 1990. Earlier records are far less complete, but they document gradually declining temperatures back to 1000, with 1000 to 1100 being about 0.2 degrees F (0.1 degree C) above the millennial mean. Earlier than 1000, the record sputters, because we lack good proxy sequences.[14] It appears that Hubert Lamb was correct, at least as far as Europe was concerned. The eleventh and twelfth centuries, and perhaps the preceding two centuries, were relatively warm and settled,

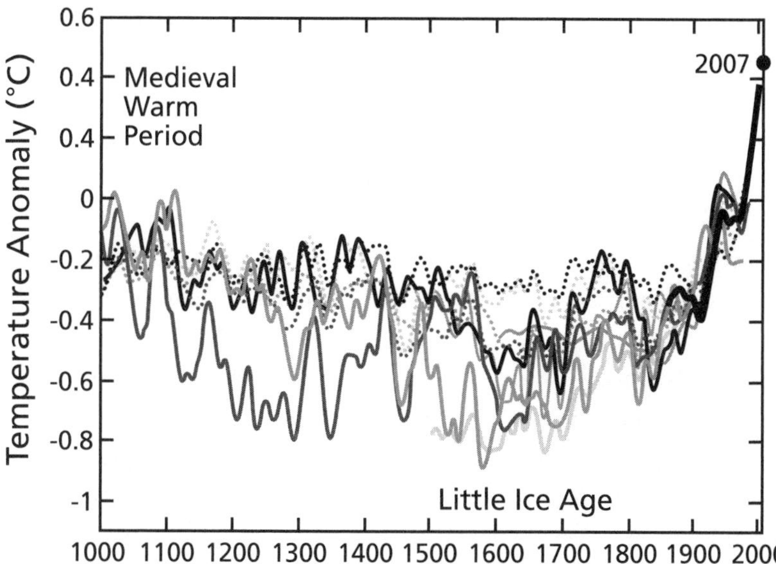

Reconstructions of northern hemisphere temperatures, compiled from the work of six different research teams. These are combined with the global mean surface temperature record obtained from instruments (shown in dark). Each curve is somewhat different; each is subject to different uncertainties and limitations that increase as one goes back in time. But the reconstructions are generally consistent over the past 1,100 years, and especially for the past four centuries and the past 150 years with their warming. The temperature fluctuations of the Medieval Warm Period reflect constantly changing climate conditions. (From the National Research Council's Surface Temperature Reconstructions of the Past 2,000 Years *[Washington, D.C.: National Academies Press, 2006], fig. S-1. Details of the different temperature curves, which are irrelevant here, can be found in that publication.)*

but with temperatures slightly cooler than those of today. Few scientists doubt that today's persistent warming is humanly caused and unique.

But was the Medieval Warm Period a global phenomenon, marked by universal warming and generally benign climatic conditions? Unlike the Little Ice Age, which left a significant climatic imprint in lands as far apart as New Zealand, the Andes, and Greenland, the warmer medieval centuries were more elusive in their impact. Then, as now, all climate was local, even if it originated in much larger-scale interactions between atmosphere and ocean. Lengthy cycles of warmer conditions in Europe brought a measure of stability to food supplies and produced favorable conditions that fostered the development of larger, more powerful kingdoms. In contrast, the same centuries gave birth to episodes of sometimes catastrophic rainfall and intense droughts to people living in arid and semiarid lands: in western North America, in India, on the margins of deserts—for example, the Sahara—and on the Eurasian steppes, where water supplies varied widely. The eastern Pacific was cool and dry; the Arctic saw much less summer ice. The term "Medieval Warm Period" is something of a misnomer, but most people continue to use it on the grounds that everyone knows what centuries are involved and because, as we shall see, there is at least sketchy evidence for warmer temperatures from Tibet to the Andes, western Europe, and North America to tropical Africa. The Medieval Warm Period is some form of global phenomenon, though not quite the one Hubert Lamb originally envisaged a half century ago. But, beyond doubt, these warmer centuries brought enormous benefits to a Europe basking in summer warmth and good harvests, especially between 1100 and 1300, during the High Middle Ages. That warmer, more stable climate lasted but two hundred to three hundred years, yet this was long enough to transform history.

A CACOPHONY ASSAULTS the ear on every side. People jostle around the stands, bargaining, gossiping, and handling produce. Brightly colored lettuces and carrots lie on piled market stands. Women sniff suspiciously at ripe apples. Farmers in tunics and stockings quaff wooden tankards of ale in the shade. Suddenly, silence falls. The crowds part as a procession of men-at-arms in bright regalia escort the lord of the nearby castle through the market. He rides a fine white horse, ornately caparisoned; wearing light armor and steel helmet, he looks neither left nor right. The silent townspeople touch their forelocks or go down on one knee. His lordship nods solemnly and passes on, his squires and attendants riding with him in close order. As the procession departs, the bustle resumes.

A.D.	EUROPE	EURASIA	SAHARA / SAHEL	ARCTIC / NORTH AMERICA	WESTERN NORTH AMERICA
				c. 1450. Norse abandon Greenland	
1400	Little Ice Age		↑		wetter ↑
	wet; 7-year famine		cooling		prolonged drought
1300	cooler, unpredictable conditions	warmer		deteriorating ice conditions in the north	
1200			many droughts		
	warmer, perhaps 1 to 2 deg. F			favorable ice conditions off Iceland and Greenland	drought
1100		cooler intervals	colder spike		wetter interval
			unpredictable conditions		
1000	unpredictable, often warm summers		many prolonged droughts		
900					drought; low lake levels
800		warm and dry (Sol Dav)	cool spike		
700	cooler and wetter	cooler and wetter	cooler	cooler	cooler

A.D.	WESTERN NORTH AMERICA	MAYA CIVILIZATION	ANDES	PACIFIC	CHINA	INDIAN OCEAN
	↑		↑	↑		
1400	wetter		warmer, wetter	warmer, wetter	drought on Tibetan Plateau	drought in Southeast Asia
1300	great drought in northern Southwest	↑ wetter, but some droughts	Quelccaya drought	New Zealand warmer		
1200			prolonged, cold, dry conditions		eastern Chinese temperatures above long-term mean	strong wet monsoons
1100	"Chaco drought"			prolonged cold, dry conditions (southern California, Palmyra Island, and from 992-1091 in New Zealand)		
	wetter interval					
1000		drought				low lake levels in East Africa. Nile floods generally low
900	drought; low lake levels	drought			Huguangyan drought (South China)	
800		drought				
700	cooler	drought				

Climatic trends across the world during the warm centuries. This is a very generalized table for guidance only.

The lord might appear in all his splendor, escorted by uniformed retainers and soldiers; endemic warfare might consume the energies of kings and barons. But behind the façade of princely processions and splendid display lay a continent that lived with the constant threat of hunger. The margin between plenty and starvation was narrow indeed, defined by unexpected spring frosts, long weeks of heavy rain, or months of seemingly unending drought. Everyone living in the countryside suffered through periods of malnutrition. We know this from the telltale stress lines found on their bones, marks of suffering they took to their graves. Even in good years, many rural communities survived at the subsistence level, or close to it. All it took was a period of heavy rain, some floods, or an epidemic of cattle disease to bring hunger to the threshold. Even in the best of times, farming was unrelentingly hard labor. The life expectancy of a Winchester farmworker in 1245 was about twenty-four years—if he survived childhood diseases. (If one factors in the high infant mortality rate, life expectancy was even shorter.) Occupational conditions such as spinal deformations from lifting heavy bags or scything hay are common among the dead found in medieval cemeteries. Fisherfolk suffered from osteoarthritis of the spine from moving boats and hauling laden herring nets. The human cost in constant, backbreaking toil and inadequate diet was enormous, even in good years.

The warm centuries brought significant relief to Europe's subsistence farmers. The growing season for cereals was as much as three weeks longer. Summer after summer, warm, settled weather would begin in June and extend through July and August into the hectic days of harvest. Even more important, the May frosts that had plagued growing crops for centuries were virtually unknown between 1100 and 1300. Warm summers and mild winters allowed people to take risks with planting marginal lands and at higher altitudes where, hitherto, colder temperatures would have precluded any form of cultivation. A growing farming population moved northward and uphill.

The figures speak for themselves. Small communities of farmers flourished at 1,049 feet (320 meters) above sea level on Dartmoor in southwestern England during the twelfth century. None farmed there in the twentieth century. Today, the Pennine Moors in northern England support no crops; but in 1300, the local shepherds complained about encroaching farmlands. Kelso Abbey, in southern Scotland, had well over 250 acres (about 100 hectares) under cultivation at an altitude of over 980 feet (300 meters) above sea level, well above today's limits. Fourteen hundred sheep and sixteen shepherds' households thrived on the abbey's land. Farmers grew wheat as far north as Trondheim in Norway.[15] Far to the south in the Swiss Alps, smallholders planted crops deep in mountain river valleys that had been covered with glaciers two centuries earlier. At lower altitudes, longer growing seasons reduced the risk of crop failure significantly, while the warm weeks of the summer growing period increased yields and allowed the accumulation of at least some food surpluses, which fed growing towns and cities. Herds grew; rural and urban populations increased. The demand for arable land skyrocketed as the calls of the church and the nobility on commoners for labor, taxes, and tithes increased steadily. Europe rang with the sound of iron axes striking down primordial oak forests and clearing new land.

CHAPTER 2

"The Mantle of the Poor"

Most pleasant farms have obliterated all traces of what was once dreary and dangerous wastes; cultivated fields have subdued forests; flocks and birds have expelled wild beasts; sandy deserts are sown . . . and where once was hardly solitary cottages, there are now large cities.

—Tertullian, *De Anima* (second century A.D.)[1]

THE PLOWBOY CRACKS HIS GOAD AGAINST his oxen's flanks in the gloom of early morning. The beasts lower their heads and strain at the harness ropes, their hooves churning the glutinous soil that glistens in the wet. At the rear, the plowman, ankle deep in mud, grasps the handles of the wheeled plow, pressing down, guiding the blade into the heavy soil. He gasps with the effort, presses down again, then lifts as the moldboard laboriously turns aside a deep furrow. Slowly the plow moves ahead along the narrow strip, following a course parallel to the furrows plowed the day before. The blade stops and starts as it slides, then balks in the clumped earth. The boy shivers in the cold as he shouts and prods the oxen to keep them moving ahead.

By late afternoon, the plowing for the day is done. The plowboy dismantles the plow and leads the oxen back to the village. He fills their manger with hay and carries out the fresh dung for later use as fertilizer. Tomorrow, the plowing will begin again, the yoking of the beasts, the assembling of the plow, the laboring across the fields.

This harsh routine played out for centuries, from England and Scandinavia to the south of France, from Spain to central Europe. A thousand years ago, Europe was a rural continent, a place of growing towns and burgeoning cities, but still a place where most people survived off subsistence agriculture and the margins between hunger and plenty were thin. Good harvests meant everything to the countryside, and it was in the countryside that the greatest impact of the warm centuries came to be felt. Each village, each town, lived from season to season. The contrast between summer and winter was stronger then. Precious summer months were full of light and wide, sunny skies, a time of warmth, planting, then harvest. These were the days of exuberance and festivals, of comparative plenty. Winters were dark and cold, with short days and muted colors, the fields bare, trees without green foliage. The season of darkness passed without electric light or gas flames, lit at best by flickering candles and warmed only by smoky fires and the great hearths in the palaces and castles of the nobility. People slept huddled together for warmth. A warm robe and a comfortable bed were prized luxuries. The warm centuries brought significant relief from the stark contrasts of the seasons. They also stimulated population growth and violence.

VIOLENCE WAS A fact of life in medieval Europe and an integral part of politics.[2] Assassinations, betrayals, fleeting alliances, and brutal military campaigns were part and parcel of existence for the elite and the privileged. Knights and the more powerful members of society placed great emphasis on displays of courage and power. Jousts tested individual bravery and prowess with the lance. Confrontations between rival landowners did not necessarily involve vast casualty lists. Often they were a way of defining territorial and political boundaries, of assessing the limits of authority and establishing who could exploit whom. Some of the campaigns were virtually ritual displays. In a time of generally good harvests, at least small-scale warfare erupted somewhere most summers.

Strong forces of disunity made it hard to forge larger political entities. Francia, the kingdom of France, was little more than an abstract concept during the early eleventh century, a series of self-governing entities that engaged in constant and often bitter rivalry. In France, royal authority extended only to the lands the king could tax and exploit. The early Capetian kings, whose dynasty began in 987, ruled not by birthright but because of their individual abilities. They created an ideology that proclaimed they were chosen by God, and they spent most of the eleventh and twelfth centuries bringing their rivals—barons and rich landowners with their strongly fortified castles—into submission. They did so in part in the name of the Church, whose churches and monasteries were favorite targets for predatory lords.

The point and counterpoint of violence ebbed and flowed during the warm centuries. The lay lords, episcopal princes, and religious communities that controlled the lush countryside, with its vineyards, plentiful harvests, and large flocks, were attractive targets for brigands and ambitious landowners intent on annexation. Few villages and small towns had much protection against marauders at a time when economic activity and rural populations were expanding rapidly. Some parts of France, such as Brittany, were in shambles from vicious strife driven by greedy neighbors who cast covetous eyes on high-yielding agricultural land. Only the western, Celtic-speaking regions escaped invasion, for they were far from fertile, with most of the population concentrated in fishing villages along the coast.

Warfare fed on food surpluses—on the ability of ambitious lords to feed their armies and to finance the construction of the stone castles that served as staging posts and bases for suppressing rebellion. Control over rich farmlands and their harvests was achieved by political marriages and by brute force. Nevertheless, some regions, such as Normandy in the north, with its dairy farms and plentiful harvests, achieved considerable stability. The Duchy of Burgundy boasted fertile lands, small farms, and larger estates; it was notably prosperous during the warm centuries, thanks especially to a burgeoning wine trade that satisfied thirsts as far away as northern Spain and England.

With rapid population growth and a growing volume of long-distance trade, shifting rivalries and alliances between great lords marked the political landscape. In what is now France, the quest for power ultimately revolved around the rich agricultural potential of the north. The warm centuries brought abundant harvests to a region known for its grain, fruit, and wine. Ample, well-watered pastureland brought massive increases in flock populations and much higher wool production. Nonfood crops, such as flax for linen and woad for dye, took up even more agricultural land. The extensive woodlands of the north offered pasturage for pigs, but were also heavily exploited for lumber, for firewood, and for charcoal used in iron production. The chronic warfare of the day stimulated craft industries in weapons manufacture and armor fabrication, while the same skills could produce axes, plowshares, and other implements of tillage in peacetime. But princely rivalries, destructive to village life and disruptive of agricultural production, prevented a spectacular economic transformation.

Warfare was endemic in the warm centuries, but gradually the Capetian kings asserted their authority and forged a true kingdom out of chaos. King Philip II Augustus made Paris the capital of France in 1194. Normandy was annexed in 1204; by 1249, most of southern France was under royal control. Thanks to clever administration, the elites of conquered lands developed a vested interest in the success of the kingdom, the crown being the icon of French unity. Thus, the warm centuries, with their dependable harvests that stimulated both trade and warfare, also witnessed the beginnings of modern Europe.

Europe overflowed with energy during the warm centuries. The art historian Kenneth Clark puts it well: "It was like a Russian spring. In every branch of life—action, philosophy, organization, technology—there was an extraordinary outpouring of energy and intensification of existence."[3] The deeds of monarchs and princes meant little to most Europeans, whether free or enslaved, the latter perhaps some 10 percent of the population in some areas. Europe was a rural continent at the beginning of the warm centuries: most people's lives revolved around the hamlet, the village, and the endless routine of planting and harvest. Despite high infant and child mortality, numerous deaths in childbirth, frequent epidemics, and occasional, persistent famines, population soared. Between the year 1000 and the outbreak of the Black Death in 1347, the population of the continent rose from about 35 million to about 80 million people.[4] What is now France had about 5 million inhabitants in the year 1000, as many as 19 million by 1350. Italy's numbers rose from 5 million to about 10 million, England's from around 2 million to about 5 million. Such growth occurred throughout Europe, albeit at different rates, with as many as half a million people living in Norway by 1300, where arable lands were in short supply owing to a relatively brief growing season. Rapid population growth during centuries with relatively favorable climatic conditions but a finite supply of good agricultural land led to an inconspicuous but widening gap between an expanding population to be fed and a shrinking area of land to feed it. The figures for England alone are daunting. In 1000, an area of some 8.5 million arable acres (3.4 million hectares) under cereals and other crops fed as many as 2.5 million people. After three centuries of population growth during long periods of good farming conditions, an expanded acreage of 11.5 million (4.6 million hectares), much of it carved out of marginal agricultural land, had trouble feeding 5 million.[5] These statistics may be misleading, for some manors engaged in intensive agriculture, especially those supplying growing cities or the grain trade. But rapid population growth, fueled in part by favorable climatic conditions, created challenging problems for the subsistence farmer.

Gothic cathedrals, illuminated manuscripts, ethereal woodwork—the material achievements of the High Middle Ages all depended on abundant food surpluses produced by the anonymous labor of subsistence farmers. These food surpluses generated wealth and money for wages to pay artisans and nonfarmers, as well as the means to honor the Lord. When harvests were abundant and life was good, both noble and commoner gave thanks and endowed God with lavish gifts, lest he unleash his wrath in the form of plague, war, and famine. In lean years, the gifts dried up and cathedral building slowed. Despite years of good harvests, the realities of plenty and hunger define the warm centuries when medieval Europe prospered and became the precursor of a continent of sovereign states.

TORRENTIAL RAIN TURNING to sleet lashes the village, turning muddy paths into small rivers. Savage gusts of wind tear limbs from the bare trees. The relentless gale shrieks through hedgerows and over thatched roofs, tumbling gray clouds across the sky, shredding the wood smoke that seeps from chimney and rooftop. No one is in the open. The cluster of dwellings seems to hug the ground, cowering from the squalls. Inside, the roar of the storm is muted, but one can barely see for the choking smoke from the hearth that hovers above the beams. Powerful scents assault the nostrils—cow dung, human sweat, decaying food, excrement. Everyone huddles silently by the warmth, wrapped in sheepskins and leggings. Cattle shift restlessly in the byre at one end of the house. Human and beast alike are waiting for a break in the storm.

Even in the warmest of decades, the climate of medieval Europe was one of sharp extremes. Weeks of snow, epochal winter storms, powerful tempest-driven surges in the North Sea, long summer droughts: subsistence farming was a challenging enterprise in even the warmest of years. With unpredictable temperatures and rainfall, medieval farmers were conservative at the best of times, just as their counterparts in developing areas are today. When one lives with the specter of hunger, one tends to hedge one's bets. Innovation of any kind can wither on the vine when faced by the cautious opposition of public opinion. Consensus in subsistence communities—based, as it so often is, on collective experience over many years—is the bedrock of survival. This made the selection of dates for planting and harvests for cereals, vines, and other crops a matter of careful deliberation, even in warmer times when crop yields tended to be higher. Rising population densities and generally favorable climatic conditions seriously challenged the conservatism of medieval farmers. Major innovations in agricultural methods took hold very widely during the warm centuries in the face of land shortages and more mouths to feed.

The milder winters, warm summers, and longer growing seasons of the Medieval Warm Period were a powerful catalyst for steady population growth, fueled by good harvests. As rural populations rose, so the demand for light, well-drained, and easily cultivable soils exceeded the supply. The soft earth of such land could be scratched effectively with the lightweight ard, a plow design unchanged since pre-Roman times a thousand years earlier. The ard is basically a scratch plow that cuts a shallow furrow through the soil but doesn't turn the sod. Medieval farmers used oxen to pull their ards, or, if they had no animals, a husband-and-wife team would plow, the one pulling, the other guiding the shoe. As long as the soils were relatively light, the ard was a simple way of furrowing a field; it had been in common use for at least four thousand years.

But ards had serious limitations. They were much less effective in heavier, more clayey conditions, where the topsoil was much harder to turn over, especially in dry periods, when the soil baked hard. Prolonged dry spells like those of the warm centuries militated against easy plowing. As the demand for arable land accelerated, so farmers moved on to these damp and often densely wooded soils, which were potentially productive but difficult to cultivate.[6] Fortunately, a new design of plow appeared during the seventh century or thereabouts, just in time for the warm centuries and more effective in heavier soils. The moldboard plow had sharp blades, which cut the soil, and angled moldboards to turn it over, burying weeds and turning up the nutrients. Teams of oxen usually dragged these plows until someone on the Continent developed a horse harness with a rigid collar that rested on the shoulders. Once horses could be used, they provided four- or fivefold the pulling power of oxen. Horses also plowed twice as fast, but a team of four of them, or eight oxen, was an expensive proposition for a single plow, although

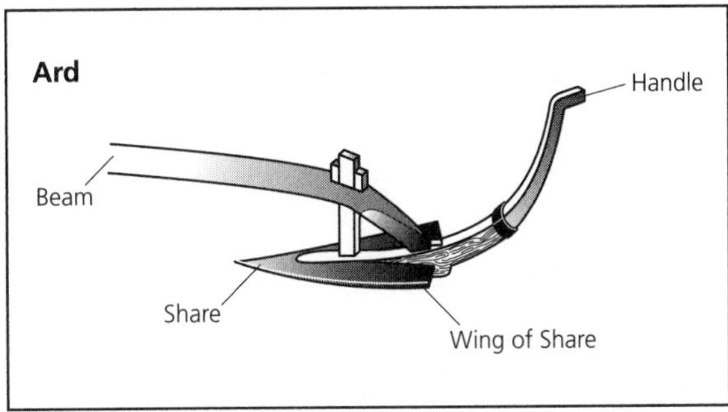

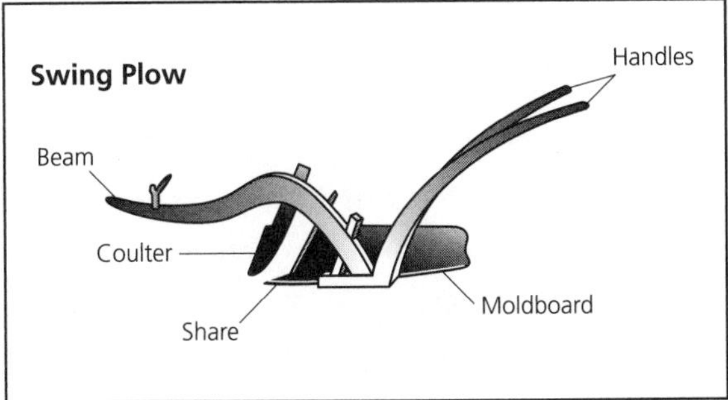

An ard in its most refined form (top) *and a simple swing plow with moldboard* (bottom).

viable for religious houses and manor estates. Village farmers solved the problem by pooling their beasts during plowing season. Even with oxen, plowing was brutal dawn-till-dusk work.

Horses and wheeled plows came into widespread use at about the same time as the three-field crop rotation system, which first appeared on monastic lands in northeastern France during the ninth century and spread gradually throughout Europe. Originally, a village would have had half its fields under cultivation at one time. Now the villagers cultivated two thirds, with only one field in three lying fallow. Three-field systems produced more grain and fodder for animals, better nutrition, larger families, and more draft animals—provided there were enough people to carry out the extra work of planting and harvest, also more plows, harnesses, and yokes, and other equipment that required carpenters and blacksmiths, wheelwrights and other specialists.

The three-field system involved planting one field with winter wheat, barley, or rye. A second field came into use in spring with oats, chick-peas, peas, lentils, or broad beans. The third field remained fallow. The three-field system spread labor requirements more evenly over the year, and provided oats for feeding horses. Legumes, such as peas and beans, fixed nitrogen in the soil and maintained its fertility, so one could keep more animals. The risk of famine was significantly reduced, while additional manure fertilized the soil. As protein intake rose, so nutrition and health improved and the population grew. Most important of all, crop surpluses increased considerably, both because of warmer climatic conditions and better harvests and because of more intensive farming.

Crop yields from medieval village farming were low. Here the pace of innovation was slowest. But they were much higher in some regions, notably in the Low Countries and northern France, as well as in parts of southeastern and eastern England, where manors and farms supplied growing urban markets or ships carrying grain overseas. Intensive mixed farming on some fourteenth-century estates in Norfolk produced yields as high as 15 to 25 bushels an acre (5.28 to 8.8 hectoliters) or more, yields normally associated with the intensive and highly efficient farming methods introduced to England in the eighteenth century.[7] These kinds of mixed farming estates combined agriculture with cattle and especially sheep farming, producing yields two thirds greater than those from well-managed manor lands in places like Winchester in southern England. The reason for this productivity isn't hard to find: the need to feed growing urban populations in towns and cities.

THE NUMBER OF towns increased exponentially between 1000 and 1400. In central Europe alone, fifteen hundred new towns appeared from the eleventh century up to 1250, and a further fifty more in the following half century. Towns varied enormously in size, some being little more than large villages, others communities of two thousand to three thousand people. All medieval towns were more densely settled than villages. There were more artisans, too, people practicing specialized trades: blacksmiths, potters, weavers, wheelwrights, and many others. Every town had a regular market; no town could have survived without one. Some even had mints, for town markets were based not on barter but on coinage. There were sometimes imposing public buildings, such as large churches and market halls, as well as other structures. Above all, towns were bustling places where there were "traffic jams." William Chester Jordan lists some of the causes in medieval Southwark outside London: "The hustle and bustle of ox carts jostling against carriages, long lines of wagons bringing fruits and vegetables, raw materials and finished goods to markets, artisans' shops and warehouses, and the seemingly endless cavalcade of mounted men and women bearing messages, coming to shop, visit or attend meetings and clamoring at others on the street to make way." He adds: "No traffic jams, no town."[8]

At first, the towns were politically weak compared with the lords of the countryside, who often controlled them. Their agents vied with the clergy for political control, but eventually merchants and the commercial sector assumed increasing power and influence thanks to an explosion in trading activity of all kinds. In an era before good roads, most goods traveled up and down rivers and canals and along seacoasts. Such trade went back to before Roman times. By the ninth century, King Charlemagne controlled important trade routes across the North Sea. The Meuse, the Scheldt, and the Rhine penetrated the heart of Europe and reached the coast in low-lying Flanders, where growing trading towns like Bruges, Ghent, and Ypres soon became crowded, prosperous cities. Other major entrepôts developed in addition to ancient centers like London or Paris: Southampton in southern England, sheltered by the Isle of Wight; Dieppe, a center of wine trade and of the herring fishery; and places like Bergen, with its cod warehouses, and Rostock on the Baltic, with its connections to the Vistula River and Eurasia.

The growth of cities had lasting political and economic consequences. In England alone, there were at least sixteen cities with ten thousand or more inhabitants by 1300; these large and ever growing numbers of people depended on others to feed them. Medieval cities became magnets not only for artisans and well-connected merchants, but also for the poor and the landless, who lived in crowded tenements and hovels either inside or outside the city walls. Like towns, the new cities were busy, crowded places, teeming with the destitute and the merely poor, with all the volatile potential for social unrest that this implied. In times of famine, hunger fueled anger in congested slums, where violence lurked close below the surface. The warm centuries enjoyed generally plentiful harvests, but the cities that mushroomed in their train were increasingly vulnerable to poor harvests. In later times, the greatest fear of England's Tudor monarchs was urban unrest caused by grain shortages.

The population changed rapidly through the Medieval Warm Period. At its end, London had perhaps 80,000 to 100,000 people, a population that made it second only to Paris among cities north of the Alps. A major sea and river port, the city drew on an irregular area of about 4,000 square miles (10,350 square kilometers) for its grain supplies, some of it from estates more than 100 miles (160 kilometers) away. Growing cities with burgeoning populations depended on reliable food supplies. When serious famine descended on England in 1315–17, the king ordered his sheriffs to procure essential provisions and hay for his household from as far away as Sussex, Cambridgeshire, Norfolk, and Gloucestershire, as much as 150 miles (240 kilometers) away. London's hinterland expanded and contracted with climatic conditions and the harvests that resulted from them. During the best decades of the warm centuries, the capital drew on perhaps a fifth of England's total arable land, of which less than half provided grain regularly. Cities like Winchester had much smaller hinterlands, in the case of the latter possibly extending no more than 12 miles (19 kilometers) away. In a world where perhaps 10.5 percent of England's population dwelled in towns and cities—that is, some 420,000 people—the agricultural requirements of urban markets were still restricted and selective. But this was to change in the future, as cities grew and climatic conditions became more unpredictable.

The real impact of more favorable agricultural conditions and longer growing seasons was in the countryside. And there, as villages grew and lighter soils were taken up, food shortages became a reality. The logical solution was to take up more land, acreage hitherto off limits because of heavier soils, marshes, hillsides, and higher ground, some of it marginal because it was subject to erosion in heavy rainstorms. But most of the new arable came from clearing a landscape covered by forest since the Ice Age.

THE SCALE OF deforestation during the warm centuries is mind-boggling.[9] In A.D. 500, perhaps four fifths of temperate western and central Europe lay under forests and swamps. Half or even less of that coverage remained by 1200, and most of that clearing took place during the Medieval Warm Period in a massive onslaught on the environment. In the Netherlands, farmers reclaimed land from the North Sea by what has been called "offensive dyking," which turned small islands in the coastal archipelagoes into larger ones.[10] Huge areas of peat moor behind the coast were drained with ditches, then reclaimed laboriously in a cat-and-mouse battle with the water as the peat subsided. At first the reclaimers depended on ever deeper ditches; later, they relied on pumps and wind or water mills. The labor involved was enormous and continued for many generations, but, in the end, the moors became pastureland for sheep and cattle and arable for crops.

Stripping Europe of its primordial forest was an act thick with cultural, economic, and political overtones. The farmers who cleared the forest deprived themselves of the safety net that a Scandinavian proverb called "the mantle of the poor."[11] Forests provided building materials, timber, firewood and game, medicinal plants and food, and browse and grazing for farm animals. The medieval farmer used more iron than ever before for axes, plows, and weapons—the metal smelted with charcoal from the forest. Great trees provided timber for cathedrals and palaces, for ships and humble structures like mills. Water mills were the new machinery of the age, as were windmills constructed almost entirely of wood. There was so much demand for timber for windmill vanes in Northamptonshire, England, in 1322 that complaints arose about deforestation. By the twelfth century, forest use was subject to intricate regulations that covered everything from grazing rights to firewood collection. Many different stakeholders, including the crown and the nobility, as well as humble folk, had rights in the forest, such as the right to hunt, to graze animals, and to use clearings. For example, many English peasants had the right to acquire construction timber, and firewood, deadwood that was knocked or pulled off trees, "by hook or by crook." The dense trees and undergrowth were a means for survival. Increasingly complex regulations surrounded the forest and the right to use and clear it, which involved balancing royal privileges and landowners' rights against the long-established economic needs of peasants.

Dark forests were a complex presence in medieval life, with many uses and powerful symbolic importance, places where powerful forces lurked and great animals like the fierce aurochs, the long-horned wild ox, thrived. The forest was also the site of the royal and noble hunt, an activity reserved for the aristocracy that signified far more than the mere acquisition of meat. The hunt was a ritual display of courtly ceremony and power, even an enactment of the conquest of wilderness by the taming of wild beasts. The Metropolitan Museum of Art in New York owns seven medieval tapestries, *The Hunt of the Unicorn*, woven between 1495 and 1505, that commemorate centuries of hunting ritual. They show a symbolic medieval hunt: the hounds unleashed; the finding of the unicorn in its hiding place; the pursuit; the mythic beast at bay; and then the kill with the huntsman's sword. The unicorn is an imaginary creature, and the tapestries show an idealized image of the hunt, but they convey its elaborate, ceremonial nature. The link between the royal presence and the overcoming of nature was irresistible, so inevitable conflicts arose between the nobility, who wanted to preserve the forests for hunting, and the rest of society, which valued the products of forested land. In the end, agriculture prevailed. The primordial forest shrank rapidly during the warm centuries in ferment of change and experimentation, and because of the intensified agriculture created by growing urban centers, higher population densities, and more mouths to feed.

After the collapse of the Roman empire in the west in A.D. 476, Germanic tribes such as the Burgundians and Vandals from east of the

Rhine overran most of what had been Gaul. The invaders arrived, perhaps eighty thousand of them, at a time when western Europe's population had declined by some 40 percent, from a peak of about twenty-six million in Roman times, as a result of plague and famine during the sixth century. The newcomers simply filled out the existing cleared land, much of which had been abandoned. To the Germanic tribes, the forest, a sacred entity to be left alone, was impassable and protected. So the period of serious deforestation did not begin until the tenth century, when economic expansion and population growth in western Europe were compounded by migration from Scandinavia and central Europe.

The surge in forest clearance came during the Medieval Warm Period, at a time when rainfall diminished by as much as 10 percent and temperatures rose between 0.9 and 1.8 degrees F (0.5 and 1 degree C). As local populations rose, so people took up abandoned or neglected lands. It is possible that the availability of new tracts that required only clearing led to earlier marriage, an increase in the birthrate, and perhaps larger families. Constant warfare entailed hungry armies, and the rising demands of a now more powerful church compounded the food supply problem. Confronted with potential shortages, growing families had two choices: they could lessen the number of years they allowed some of their fields to lie fallow (a dangerous practice in the long run, because of catastrophically lower crop yields from more readily exhausted land), or they could clear new fields. Fortunately, there was plenty of land to go around, so people often carved new acreage from the margins of the forest, a process known as assarting. ("Assart" comes from the French word *essarter*, "to clear." It also implies the uprooting of stumps, an essential part of such clearing.)

Day after day, men with axes would climb high in the trees, lopping off branches that were then piled up and burned. Sometimes a young man would lose his balance and fall, landing with a thud on the hard ground. Maybe he would be lucky and escape with bruises. More often, he would break an arm or leg and perhaps be crippled for life, or at best be one more mouth to feed until he could fend for himself again. Once the lopping was over, the tall trunks would stand bare and stark at the edge of the receding forest, to be chopped down by strong men working in unison, pausing frequently to sharpen their axes. Back at the nearby village, a blacksmith would be hard at work, hammering bent blades and forging new axes to keep up with demand. Slowly a new field would emerge from the forest, studded with large stumps. Once the trees were gone and the branches burned to fertilize the land with the ash, the villagers would move in and laboriously cut and pull the stumps from the ground, using oxen or horses to help them with stout ropes and iron chains.[12]

Assarting was laborious and labor-intensive, usually a prolonged process. It began with periodic burning to clear brush, and heavier grazing of surrounding woodland. Eventually the forest deteriorated, at which point the assarters moved in, cleared the stumps, and founded new villages. Completely new settlements often rose in remote clearings in the forest, especially those of monastic houses that sought seclusion in the wilderness.

The warm centuries saw thousands of new settlements throughout western Europe. In France's middle Yonne Valley, southeast of Paris, lords encouraged settlement and clearance by granting self-government to those who cleared new land and reducing or abolishing customary taxes on them. Payments in labor were waived: peasants were allowed by their controlling lords to marry outside their own community. As the French historian Marc Bloch once remarked, a kind of "megalomaniac intoxication" gripped many proprietors with grandiose visions of new landscapes where wastelands had become profitable acreage that would yield more wealth in kind and ease population pressure on farming land. In the east, German lords, both lay and ecclesiastical, encouraged colonizers to take up forested and swampy lands east of Berlin where small bands of elusive hunters lived. In the words of one appeal for recruits: "These pagans are the worst of men but their land is the best, with meat, honey, and flour. If it is cultivated the produce of the land will be such that none other can compare with it."[13]

One can argue for a direct connection between medieval warming, population growth, and farming innovation, but there were underlying social and religious factors as well. In earlier times, manor owners tried to keep peasants confined to their lands so they could control them and accumulate more rent. As the warming came, local rulers and church leaders in an increasingly pious age made constant efforts to consolidate their control for political gain and also to accumulate wealth. They gradually assumed the right to dispose of uncleared, unused land as if they were rulers, granting wilderness to groups of colonists who would clear and farm it, "bring it into the realm of human affairs."[14] People and their labor became a source of wealth for lords. These colonizers soon became free farmers, who owned land and could earn money for their crops. With the emancipation of the commoner and the widespread adoption of primogeniture, younger sons needed outlets—new land. Forest clearance was to them what the Crusades and wars of conquest were to the nobility.

The religious played a huge role in forest clearance and the agricultural revolution. The Benedictines, in particular, considered manual labor as important as reading or prayer. Work had spiritual rewards. Saint Bernard wrote: "A wild spot, not hallowed by prayer and asceticism and which is not the scene of any spiritual life is, as it were, in a state of original sin. But once it has become fertile and purposeful, it takes on the utmost significance."[15] Benedictine communities did much to dispel the ancient dread of primordial forest among medieval peasants. The historian Michael Williams calls religious orders the "shock troops" of forest clearance. Numbers tell the story. Between 1098 and 1675, the Cistercians alone founded 742 communities, 95 percent of which were in existence by 1351. Each house engaged in intensive farming and forest clearance. Wrote Gerald of Barri: "Give these monks a naked moor or a wild wood, then let a few years pass away and you will find not only beautiful churches, but dwellings of men built around them."[16]

By any standards, the deforestation of Europe during the warm centuries ranks as one of the greatest such episodes in history. France's forests were reduced from 74 million to 32 million acres (30 million to 13 million hectares) between about 800 and 1300, but a quarter of the country was still forested. Overall, perhaps more than half of Europe's forests were cut down between 1100 and 1350. Clearance in Britain was more piecemeal, with less planned deforestation. Even so, the statistics of population increase are impressive by any standards. In just one tiny English parish, Hanbury in northeast Worcestershire, the population rose from 266 in 1086 to 725 by 1299.

ALONGSIDE DEFORESTATION AND agricultural innovation came a dramatic growth in sea fishing, triggered by more favorable climatic conditions offshore, Christian doctrine, and a growing demand for military rations.[17] Charlemagne and his successors had fostered maritime trade routes that connected the Baltic and North Sea coasts, partly because the Mediterranean was an Islamic sea. Most seafaring took place during the summer months, but even then the shallow waters and sluicing tides of eastern England and the Low Countries claimed many a slow-moving merchant vessel. The Medieval Warm Period brought slightly higher temperatures and longer summers to northern European waters. To judge from modern-day conditions, the warmer months would have witnessed periods of high temperatures and prolonged anticyclonic conditions when winds were calm and seas near mirrorlike. Compared with the Little Ice Age of later centuries, conditions were generally less stormy during the summer, although winter gales could still stream ashore and flood low-lying farmland, especially in the Low Countries. The warm centuries saw a rapid expansion of maritime trade that was in part a reflection of more benign conditions offshore. At the same time, sea fishing acquired growing importance, especially the traffic in salted herrings.

Clupea harengus, the Atlantic herring, is among the most prolific of all fish, massing in the North Atlantic each spring, then passing southward into the Baltic and along the Scottish and English North Sea coasts in summer and early fall. No one knows whether herring are sensitive to sea surface temperatures, but it seems likely that they are, for there have been notable fluctuations in herring shoals over the centuries that were not wholly due to overfishing. Whatever their sensitivity to sea conditions, herrings were promiscuously abundant during the Medieval

Warm Period, whereas relatively few people had eaten or fished for them in earlier times. Catching them required open boats and drift nets, also calm weather in the shallow waters of the North Sea. A combination of favorable weather conditions, especially in the early fall, and an inexorable demand for sea fish saw herring become a major international industry during the warm centuries.

Herrings are fatty fleshed. Once caught, they decay within hours unless they are salted. Salting is an ancient technology dating far back into ancient times, but crude methods like placing the fish in piles of salt were ineffective with *Clupea*. Sometime in the ninth or tenth centuries, just as sea temperatures were warming slightly, Baltic fishers developed a method of salting herring in brine packed in sealed barrels, which allowed them to harvest migrating herring by the millions. The new methods soon spread to North Sea ports. For the first time, it was possible to transport salted fish long distances inland. Almost overnight, a huge industry came into being.

Since the first days of Christianity, the devout had fasted and eaten meatless diets on holy days and during Lent. Most people ate a meatless diet of milk and grain anyhow, so the prohibition most strongly affected the noble and wealthy, as well as religious houses. The number of meatless days grew and grew. By 1200, effectively half the days of the year were classified as holy. Protein-rich fish were deemed acceptable food for such days, but there were not enough freshwater fish to go around, this despite an explosion in fish farming under the auspices of monasteries and lordly estates. Here, again, warmer conditions helped. The warm summers of the Medieval Warm Period provided ideal circumstances for farming such shallow-water fish as the carp. *Cyprinus carpio* flourishes in the wild in the muddy waters of the Danube and other southeastern European rivers. Between 1000 and 1300, carp spread rapidly across continental Europe as higher temperatures and long summers raised water temperatures in ponds and rivers. *Cyprinus* was relatively easy to farm, to the point that carp farming became a large industry in medieval Europe. But the fish were expensive. In 1356, a hundred carp served at a wedding in Namur cost twice as much as a cow. Religious houses and wealthy nobles were assiduous in cornering the market in farmed fish, rendering them effectively inaccessible to common folk.

Fortunately for the devout and for those concerned with feeding the urban poor, a cheap substitute now appeared in the herring. There was a revolution in sea-fish-eating habits during the warm centuries. Between the seventh and tenth centuries, sea fish appear only in the middens (garbage deposits) of ports and estuary towns. Bone fragments found in such ancient rubbish heaps are a mine of information about ancient diets, giving us data on such esoterica as the weight of medieval cod and the ages at which butchers usually slaughtered oxen. By 1030, herring abounded at the coastal city of Hamwic (modern Southampton) and were traded far inland. During the late eleventh century, an estimated 3,298,000 herring were landed at English ports alone, this quite apart from landings on the Continent. By the twelfth century, herring were commonplace as far from the sea as Vienna. Fish became so important to Christian observance that Pope Alexander II even permitted Sunday fishing during herring season in 1170. For six weeks every fall, a huge herring fair at Yarmouth in eastern England supplied millions of barreled fish for people on both sides of the North Sea. The sales were enormous. Kings fed their armies on salted herring. Towns paid taxes in tens of thousands of *Clupea*. In 1390 the almoner to the king of France bought 78,000 herring on the Paris market for distribution to hospices and poor households.

Salt herring was cheap and ubiquitous, but an unattractive food, spurned if possible even by the poor, for it tasted like wood unless cooked with elaborate spices. Yet the demand was insatiable. The glory days of the herring fisheries were in the early fourteenth century, during the closing decades of the warm period. It may be no coincidence that the herring schools and fish landings at places like Yarmouth suffered a major decline as conditions grew colder after 1300 and sea surface temperatures in the North Sea declined. A combination of overfishing in response to insatiable demand, better salting methods, and, perhaps, climate change tipped the medieval herring industry from sustainable to unsustainable.

The Norse introduced a much more palatable alternative in their warships and merchant vessels: dried and salted cod, often called stockfish.[18] *Gadus morhua*, the Atlantic cod, is a white fish with firm flesh that has a low fat content. Every winter, fisherfolk in the Lofoten Islands of northern Norway would catch and dry cod, which the Norse used as staple tack on their voyages. Tightly packed stacks of cod fed their crews during the great voyages of the Medieval Warm Period. During the same centuries, the Hanseatic League of the Baltic, which already had a stranglehold on much of northern Europe's grain trade and the yields of its then abundant harvests, discovered that it could generate enormous profits by trading grain from more temperate climates for cod loaded at Bergen in southern Norway. The Hanse already dominated herring commerce. They now added cod to the inventory, so English fishers started sailing far offshore, to cod fisheries off southern Iceland, where they could outflank the Hanse. At the same time, they found a substitute for fluctuating herring stocks at a time of cooling temperatures. Their fishing boats, their technology, and their experience came from centuries of fishing in North Sea waters during the benevolent summers and mild falls of the warm centuries. This apprenticeship served them well as cod fleets moved to Ireland and Iceland, and, ultimately, to the Newfoundland fisheries of later centuries.

In Europe, the warm centuries had their profoundest direct effect on common folk, the anonymous people who fed society. Kings, princes, and lords enjoyed estates and landholdings, and they engaged in intrigue, war, and sometimes crusades. Theirs was a life of elaborate ceremonies, of chivalry, cruelty, and violence. Their expertise had little relevance to the knowledge required to work the land, to cope with the vagaries of sudden climatic shifts. Most commoners worked the land; a large percentage were free laborers, many with specialized expertise. Many grew cereal crops; these people were knowledgeable about crop rotation, plant diseases, and storage. There were cowherds, shepherds, and swineherds, others who were adept at drainage. Eel catchers thrived by rivers and lakes and in remote marshes, for the easily smoked eel had become a standard currency for rent. Women raised bees, brewed ale, and spun wool. Each generation learned from its predecessor in apprenticeships and by word of mouth, passing on an enormous body of often arcane information that came with working the land in all weathers in a climate where shifts between warm and cold, wet and dry, could arrive almost overnight.

To lords and princes, the peasantry was an anonymous presence that cultivated the soil and provided food. But it was these humble folk who kept Europe fed and who adjusted their farming to warmer times. Their food surpluses fed the armies and the laborers building not only palaces but also those most impressive of all medieval structures, the Gothic cathedrals. Their labors felled the trees for the beams, quarried the stone for the towers and soaring naves. The great churches, visible for miles around, linked the living and divine worlds, for, in those pious days, all things emanated from the Kingdom of God. Great cathedrals like Sens and Chartres, built at enormous expense, were metaphorical sacrifices of stone and material goods offered in expectation of divine favor. The favor most people sought was a good harvest, for even in these warmer times, the farmer could never relax.

FOUR HUNDRED YEARS of rapid population growth and generally plentiful food supplies, of unbridled forest clearance and fast-growing towns and cities: Europe was a very different continent at the end of the warm centuries. By the late thirteenth century, however, Europe was facing serious economic problems, for population growth had outstripped the equivalent jumps in agricultural production.[19] By 1300, much of the population was worse off than it had been a century earlier, as inflation undermined wealth and the upper classes placed ever greater demands on commoners. The farmers responded by taking up marginal lands and by other shortcuts such as shortening fallow periods, which, in a time of relatively predictable summers, may have seemed logical ways of boosting crop yields. Inevitably, farmers' indebtedness to landowners increased, while economic uncertainty also struck home in cities, where the vagaries of the wool trade and other industries could wreak havoc, and military blockades were a fact of life.

All of these problems had the potential to turn much of western Europe into a powder keg, but climatic events intervened dramatically in 1315. "During this season [spring 1315] it rained most marvellously and for so long," wrote a chronicler of the day, Jean Desnouelles.[20] The rain began seven weeks after Easter, just as the new crops were in the ground. Soon, freshly plowed fields were quagmires, the seeds washed out of the ground. Muddy soil cascaded down the freshly cleared slopes of marginal lands, creating deep gullies in new fields. Corn and oat crops, heavy with moisture, barely grew ripe. The deluge continued into autumn. There was hunger by Christmas as a "dearness of wheat" began. Even military campaigning, which normally continued in any weather, stopped in its tracks. When fighting resumed, it prevented the distribution of grain to hungry communities. The rains continued in 1316, reducing many to penury as crops simply did not ripen. "There was a great failure of wine in the whole kingdom of France" as mildew attacked the grapes. The suffering continued for seven years, ending with a bone-chilling winter in 1322 that immobilized shipping over a wide area.

Grain production over much of northern Europe during these years may have fallen by as much as a third, although, of course, the effects differed from place to place. Herds and flocks were reduced by as much as 90 percent owing to diseases such as rinderpest and liver fluke brought on by wet weather. Fruit rotted on saturated trees; coastal fisheries and fish ponds were devastated, all this apart from the damage to industrial crops such as flax. At least 30 million people were at risk of malnutrition. No one knows exactly how many Europeans perished from hunger and famine-related diseases, but there were at least 1.5 million deaths, most of them the poor. Inevitably, in a pious and superstitious age, the famine was attributed to the wrath of God. Processions of flagellant penitents wound through the streets of cities and towns.

By the time the hunger years ended, the more settled climate of the warm centuries was a distant memory. Much more unpredictable conditions, greater storminess, and cycles of very cold winters or warm summers marked the gradual beginnings of the Little Ice Age. But worse was to come a little more than a generation later. In 1347, a ship from the Black Sea appears to have brought the plague to Italy. The plague bacterium, *Yersinia pestis*, lives in fleas, which live on warm-blooded hosts such as rats, humans, or cats. The Black Death, so named after the dark discolorations of the skin resulting from infection of the blood, spread rapidly along trade routes into France.[21] By 1348, it had struck England and Spain, then Scandinavia. Between 1348 and 1485, thirty-one major outbreaks descended on England alone. Eighty million people lived in Europe in 1346. Twenty-five million of them perished in the pandemic of 1347–51. From a low in 1450, England's population did not reach pre-plague levels again until about 1600, Norway's not until 1750.

There was no defense against the Black Death, which spread like wildfire among crowded urban populations. No herbal remedies worked; leeches and bloodletting provided no relief. Because no one knew how the disease was spread, nobody took measures to eliminate the rats and other animals that carried the infected fleas. Once it became clear that the risk of infection was lower away from crowded cities, the nobility and middle classes moved into the countryside. Some religious communities where monks lived in close quarters lost as many as 60 percent of their members. The church was plunged into crisis; agricultural production plummeted; landowners had difficulty finding peasants to work their land. Agricultural wages rose rapidly, but those subsistence farmers who survived may have been somewhat better off, as there was more land to go around and fewer mouths to feed, both in the countryside and in the cities. Marginal lands everywhere now reverted to fallow. But new waves of infection in the 1360s soon wiped out any gains, quite apart from the emotional impacts of the loss of a third (the proportion is controversial) of Europe's population in a few years. Only the constant wars continued, waged by rulers in the belief that the Lord had brought famine and plague on the people as punishment for their sins. Those who fought one another or practiced economic oppression believed they were doing God's bidding with virtuous acts that would restore justice to the world. Rural and urban violence rose to unprecedented heights in a Europe that, after the Black Death and the warm centuries, was never the same.

But the Medieval Warm Period had helped change Europe beyond recognition, and had widened the mental and physical boundaries of European society. During these centuries, Norse voyagers took advantage of favorable ice conditions in the North Atlantic to sail westward to Iceland, Greenland, and Labrador—lands previously outside European consciousness. These voyages will be described in chapter 6.

CHAPTER 3

The Flail of God

> *Now when he reached the Khan's camp, at the end of the third month the grass was green and trees everywhere in bloom, and the sheep and horses were well grown. But when with the Khan's permission he left, at the end of the fourth month, there was no longer a blade of grass or any vegetation.*
>
> —Ch'ang Ch'un, *The Travels of an Alchemist* (1228)[1]

"EUROPE IS SAID TO BE A THIRD OF the whole world," a fourteenth-century encyclopedist informs us. "Europe begins on the river Tanay [Don] and stretches along the Northern Ocean to the end of Spain. The east and south part rises from the sea called Pontus [the Black Sea] and is all joined to the Great Sea [the Mediterranean] and ends at the islands of Cadiz [Gibraltar]."[2] To medieval Europeans, the easterly reaches of the European plain were a virtually unknown realm, where gently rolling terrain receded into the far distance and eventually Europe became Asia. The population was sparse and, for the most part, constantly on the move, their movements dictated to a great extent by drought cycles.

Deserts and semiarid environments are extremely sensitive to even tiny variations in rainfall. An inch (25 millimeters) more rain can shrink the frontiers of a desert by hundreds of square miles. Standing water may appear where none has been present for generations. Now suppose the rains continue to be slightly higher for a few years. Pastureland appears like magic; herds of antelope feed across what was recently arid terrain. Nomads graze their sheep and cattle near water holes and wherever graze can be found. Then the rains falter. Streambeds dry up; water holes vanish; the grass withers and dies. With the experience of generations, the herders move their beasts to the margins of the desert, to better-watered lands. Arid lands like steppes and deserts act like huge pumps. When rainfall increases even slightly, the breathing desert sucks in animals, plants, and people with its promises of grazing land and water. Like a giant lung, the pump may hold its breath for a while, but, inevitably, it exhales as dry conditions return, expelling nomads, their herds, and the antelope to its margins.

Europe was well watered during the Medieval Warm Period, with slightly higher temperatures and with conditions somewhat drier but certainly favorable for subsistence agriculture. The Europeans lived on a geographical peninsula hedged in by much more arid environments, where the desert pump set the pace for much of human existence. Drought was the harsh reality of a warmer world in these environments, with the potential to change history.

The Greeks and Romans thought of the nomadic tribes of Eurasia as the barbarians who lurked outside the gates, threatening destruction, rape, and pillage. One cannot entirely blame them: the Scythian nomads north of the Black Sea were known to drink from the skulls of their victims. In the same way, medieval Europe thought of itself as surrounded by a hostile world—pressed by Islam in the east, hedged in to the west by an ocean that extended to a stormy horizon, and threatened from Eurasia by nomadic tribesmen from the endless plains. The nomad threats were real. On April 9, 1241, an army of Mongols under a general named Subutai defeated the Polish princes under Henry the Bearded at Legnica in Silesia, now part of Poland. Henry's heavily armored knights were no match for the nimble, arrow-shooting Mongols. The conquerors are said to have collected nine sackfuls of right ears from the bodies of the slain. Only the death of the Great Khan Ögötai Khan back in his homeland later the same year stopped the Mongol leader Batu Khan from pushing westward all the way to the Atlantic coast.

The Eurasian steppe is an unforgiving environment, subject to drought and torrential rainfall, savaged by extreme heat and unforgiving cold. It was the home of one of history's great conquerors: Ginghis (Genghis) Khan, whose empire expanded rapidly during the Medieval Warm Period.[3]

GINGHIS KHAN CALLED himself "the Flail of God." A brutal warrior, he descended on the settled lands of China and central Asia with bloodthirsty, hammerlike force. In 1220, he addressed the terrified citizens of Bokhara from the pulpit of the city's central mosque: "Oh People, know that you have committed great sins, and that the great ones among you have committed these sins. If you ask me what proof I have for these words, I say it is because I am the punishment of God."[4] He spoke with a shrewd knowledge of the conquered.

Like droughts and plagues, Ginghis Khan seemed an instrument of divine vengeance. Christian and Muslim alike cowered at his coming. The anonymous compilers of the *Chronicle of Novgorod*, a major staging point on the trade routes that linked Byzantium with the Baltic, called the Mongols "pagan and godless." They were ruthless in victory. "God let the pagans on us for our sins," lamented the chroniclers in 1238. "The devil rejoices at the wicked murder and bloodshed." God was punishing the city with "death of famine, or with infliction of pagans, or with drought, or with heavy rain, or with other punishment. . . . But we always turn to evil, like swine ever wallowing in the filth of our sins."[5]

Ginghis Khan played the role of a conqueror with consummate skill. He was "a man of tall stature, of vigorous build, robust in body, the hair on his face scanty and turned white, with cat's eyes, possessed of great energy, discernment, genius, and understanding, awe-inspiring, a butcher, just, resolute, an overthrower of enemies, intrepid, sanguinary and cruel."[6] While he preferred that his enemies surrender and submit, he resorted to slaughter if they defied him. When the rich Chinese city of Chung-tu, near modern-day Beijing, refused to submit, Ginghis Khan attacked in force, placing prisoners as attack troops in the frontline, then hurling the heads of the slain into the enemy's lines. A few years later, a Muslim visitor noticed a white hill near the rebuilt city: it was constituted of the weathered bones of the thousands massacred when Chung-tu fell and was burned to the ground. The greatest of all Mongolian conquerors shed rivers of blood wherever his armies campaigned.

Ginghis Khan was of humble origins, rising to prominence through sheer ability and ruthlessness. At first he was one of many leaders among a patchwork of nomadic tribes comprising some two million people and scattered across the vast Eurasian steppes. War was a way of life for the nomads, who fought on horseback. They were seasoned raiders and ruthless fighters, also fiercely independent, led by tribal chiefs who entered into alliances with others only with the objective of amassing wealth in livestock. In 1206, Ginghis Khan was elected Great Khan of the Mongols. He was a brilliant strategist and conqueror, but just as talented an administrator. He quickly broke up the ancient tribal structure, organized his army into tightly controlled standard units in multiples of ten, starting with the *harban* (ten men) and ending with the *tümen* (ten thousand soldiers). The troops fought as units; no order had to be given to more than ten people at one time. Mongol armies were famous for their ability to fire their arrows in all directions while at full gallop. Each cavalryman wore a silk shirt under mail or a thick leather garment, which protected him effectively against arrows. They wore leather or metal helmets, and they carried two composite bows made up of pieces of wood and yak horn, and at least sixty arrows. Some carried heavy lances, clubs, and scimitars; others, swords and javelins. Every fighting man carried his own provisions, cooking utensils, and other equipment in an inflatable saddlebag made from a cow's stomach. Mongol armies conquered by mobility, stratagem, and decoy tactics, and by using firecrackers made with gunpowder to terrorize a waiting army. They knew they terrified their enemies and made the most of it.

At first, Ginghis Khan's kingdom was little more than a patchwork of chiefdoms held together by the force of his personality, his military abilities, and the prospect of booty beyond imagination acquired by conquests and raids in settled lands. But, over a mere twenty years, Ginghis's armies swept across the steppes and southward with breathtaking rapidity and ruthless efficiency. The Mongols slaughtered so many human beings that the Mongolian historian Juvaini remarked that the deficiencies would never be made good on the Day of Judgment.[7] Famous cities were leveled to rubble; Iraq's irrigation system, developed over many centuries, was smashed beyond repair. Thousands of Baghdad's inhabitants were slaughtered without mercy in 1258. The ripple effects of the conquests spread far and wide. Christianity was weakened. Islam was much strengthened, but the faith that emerged from the ordeal of the conquests was narrower, more restricted, and closed to new ideas. The great traditions of Islamic learning in medicine, mathematics, history, geography, and astronomy that had flourished from 800 to 1200 now withered in the face of religious orthodoxy. Gradually, intellectual and scientific primacy passed from the Islamic world to western Europe. Meanwhile, the peaceful conditions in central Asia encouraged a few European travelers to venture along the intricate tracery of the ancient Silk Road to China, among them the Venetian Marco Polo (1254–1324), who entered the service of the Mongol emperor Kubilai Khan. By 1260, a Franciscan friar was archbishop of Beijing.

Where Ginghis Khan excelled was not only in conquest, but in his realization that an *empire*—as distinguished from a kingdom—had to be based on stable government, efficient administration, prosperous trade between the steppes and the settled lands, and law and order. He turned the expanding Mongol domains into a huge empire linked by efficient communications and kept in order by veiled military threats and the savage reputation of his troops. It was Ginghis Khan who told his armies to conquer first, then plunder, not to conquer and plunder at the same time. Rebels and chiefs convicted of treason received brutal punishment. Sometimes they were rolled in carpets and trampled by horses. Or, like one Kurdish chieftain, they were bound, smeared with sheep grease, then left to starve and be consumed by maggots.

The Great Khan considered himself the instrument of divine punishment, but, in reality, his lightning conquests owed much not only to his leadership and charisma but also to the realities of medieval climate on the steppes and to a lifeway that depended on mobility and the unique anatomy of the horse. The rhythms of nomadic life danced to the oscillations of the desert pump that brought drought, heat waves, bitter cold, and floods. These rhythms developed deep in history, long before the four centuries of medieval warming descended on the steppes. But where Ginghis Khan showed genius was in trying to move his domains away from the tyranny of the horse and the desert pump. In this he, and his successors, at least partially succeeded.

THE NOMAD'S STAGE covered an enormous tract of varied terrain from the Danube in the west in an ever widening belt that became part of the central Asian steppes east of the Volga River. The horse country extended to the Great Wall of China, more than 4,350 miles (7,000 kilometers) to the east. Popular writers often paint the steppes as a huge, rolling grassland that extends for thousands of kilometers without change. In fact, the term "steppe" encompasses a staggering variety of different environments—forest steppe, which is relatively better watered and well wooded; open grassland; river valley; marshland; and mountain range.[8] Marshes, endless forests, and open tundra marked the inhospitable northern boundaries of the steppe. To the south, grasslands and deserts run from the Nan Shan and Tien Shan mountains, the "Mountains of Heaven," in the east, along the Oxus River and the Iranian plateau, then lap against the natural bastions of the Black Sea, the Carpathians, and the Danube. But the heart of the steppe has always been the pastures along the northern edge of the Tien Shan and the southern edge of the Altai. Since the time of the Scythians, more than fifteen hundred years ago, nomadic horsemen have galloped through the low passes between these mountain chains, out of Asia into Europe.

By any standards, the distances are enormous, unfolding amidst landscapes where people and their animals are diminished to tiny dots

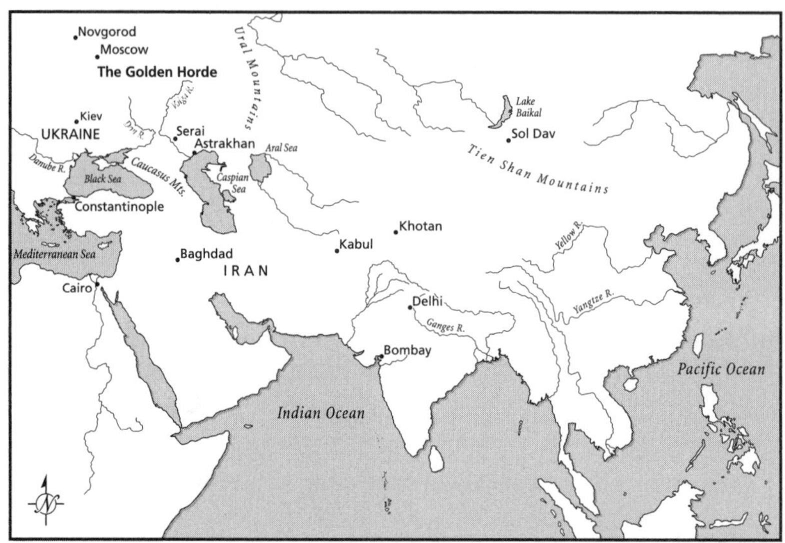

Locations mentioned in the text. Some minor places are omitted.

against land and sky. The medieval friar William of Rubreck, who came to the Mongolian court as a papal envoy in 1253–55 and visited one of Ginghis's successors, Möngke, said that the steppes were "like the ocean sea," vast, mostly uninhabited, and dangerous. To travel across such country is to be lost in the immensity of landscape, to be dwarfed to insignificance by the sheer scale of the terrain. The steppes diminish the individual. I remember once crossing the Kafue River Valley in central Africa, an utterly flat, seasonally inundated floodplain. We saw what appeared to be some trees in the distance; then they started to move. They were a herd of several thousand antelope. Ghingis Khan's horsemen would have seemed like moving trees in the immensity of the steppes, but trees that brought danger, menace, and slaughter. The Great Khan is said to have told William of Rubreck that "just as the sun spreads its rays, so my power . . . is spread everywhere."[9]

The continental climate of Eurasia is always harsh, with average temperatures falling from west to east. Flat topography, low rainfall, and frequent dry winds inhibit tree growth. Winters are eight months long, dry, often intensely cold, and windy. The endless winds redistribute the snow cover, such as it is. Summers are torrid; heat waves and droughts are commonplace. As you move eastward past the Urals, the temperatures drop even further, winter snow stays on the ground longer, and the climate is considerably drier. Throughout the steppe, plants grow deep root systems in response to the aridity, while most smaller animals live underground. In ancient times, wild horses and asses, also saiga antelope, moved across the steppe in herds of up to a thousand. They did no damage to the bunched grasses of the steppe, which thrive under moderate grazing. But overgraze the pasture, especially when the soil is wet in spring, and rapid moisture loss decimates the graze.[10]

The steppes were a demanding homeland, even for the most expert herder. Ginghis Khan could sweep aside armies and overthrow kings, but he could not control the environmental realities of the steppes or the limitations of the horses that powered his armies.

A.D. 1100. THE bitterly cold north wind of spring lashes the horsemen's faces as they huddle in their saddles, leaning into the gloom. Their compact horses trot steadily, oblivious to the chill, following a barely discernible path through the shallow valley. Beyond lie the endless reaches of arid grassland, where a man can get lost in hours with no landmarks to guide him. It is early in the year to be on the move, but the men know that survival depends on their scouting journey. They have moved north from their snug winter quarters, traveling light through the winding valley where they and their ancestors have grazed their herds for longer than they can remember. As they ride, their eyes are never still, scanning the horizon for other, perhaps hostile riders, looking for patches of good grazing and signs that rain has fallen as the snows melted. In a few days, they will ride south with the wind on their backs, armed with vital intelligence of the staples of life on the steppes—grazing grass and water. Back in winter quarters, the khan and his advisers will rely on their observations to time the band's move northward to summer pastures.

To the Mongols, the horse was everything—meat, milk, cheese, yogurt, even a source of alcohol in the form of fermented mares' milk, koumiss. Horses were wealth, prestige, a potent military weapon, and, above all, a source of freedom and mobility. By the time of the Great Khan, the compact, sturdy Mongolian steed had been an integral part of steppe life for at least 4,500 years.

Horses were domesticated in about 3500 B.C., much later than cattle, goats, and sheep, on the margins of the steppes and probably in several locales: in the Black Sea region; perhaps also in the Altai Mountains, which are still archaeologically little known.[11] Horses were much better adapted than other domesticated animals to intense cold and deep snow, and had been since the Ice Age, when they were a favorite quarry of tiny numbers of hunters on the plains. Between about 3300 and 3100 B.C., a colder, drier climatic cycle (which coincided with severe drought in Mesopotamia) led to much wider domestication of horses, which soon became central to human life on the steppe and changed history.

Horse riding was a revolutionary, if logical, shift in human transportation. It cut travel time across the steppe, allowing people to exploit widely scattered food resources, increasing territorial boundaries by a factor of five and making a mockery of earlier constraints. On the steppes, food resources could be quite rich in certain areas, such as major river valleys, but vast stretches of poor, sometimes hostile territory separated the areas of abundance. Anyone who could cover these distances relatively rapidly could survive on the steppe, and the entire shape of society changed as a result. Now one could transport large quantities of food and other goods with ease, especially when horses were combined with oxcarts. Wealth would be measured at least in part in horses; interdependency on neighbors and on settled farmers would increase, because horses were a desirable commodity thanks to which trading over long distances became much easier. Above all, mounted raiders could strike across long distances at their enemies and then retreat safely from pursuers on foot. By Ginghis Khan's time, looting, raiding, and warfare had been an integral part of plains life for thousands of years. His remote precursors, the Scythians, were the archetypal "barbarians" lurking to the north of the civilized classical world.[12] The Greek historian Herodotus wrote of their savage warfare and described how they would scalp their enemies, turning their skulls into drinking vessels, which they set in gold and hung from their belts.[13] When attacked, they would simply melt away into the vastness of the steppe.

The Scythians have been called the world's best light cavalry. It was these master horsemen who introduced horses to temperate Europe. Their successors, the Sarmatians, who broke their domination in the fourth century B.C., are said to have invented the iron stirrup, which allowed them to carry long lances while in the saddle and to push their enemies off their horses.

The steppe nomads could never settle in one place, for to do so would invite disaster from overgrazing. They lived off their herds and their horses; sometimes they planted grain at a convenient locale, then left it unattended and returned months later to harvest the crop. But they also engaged in an intricate gavotte with the climatic pendulum because of their vulnerable mounts.[14]

Horses brought mobility, but they could also be a serious liability because of their digestive inefficiency. Cattle are efficient eaters in the sense that they excrete as little as 25 percent of the protein they consume. This means that they can eat dried-out, low-protein grass and still survive. Horses digest as little as 25 percent of the protein they consume; the rest, they excrete. Both use what one can call fermentation vats to convert plant protein into energy. The cow's vat, the rumen, lies at a point in the body where food has not yet been digested. Here, bacterial action breaks down plant protein, much of it locked in the cell walls of the plant. The freshly broken-down protein passes into the duodenum, where it is broken down still further into amino acids. From there, the protein passes into the small intestine; here it is absorbed into the bloodstream and used for important purposes such as building muscles and nourishing fetuses. The horse's rumen lies in the hind gut, in a position where the food has already passed through the duodenum and small intestine. So the horse produces small quantities of amino acids and does not absorb large quantities of protein through the intestine walls. Plant protein is broken down by bacterial action in an effectively useless location and becomes nitrogen-rich protein that benefits the soil, not the animal.

Normally, graze is so plentiful on the steppes that neither cattle nor horses need to retain all the plant protein they eat. But in times of drought, plant protein is in short supply. Living grass has about 15 percent protein; dead, only 4 percent. When dry conditions kill fresh grass, protein retention becomes all-important. Cattle retain three times more protein than horses do.

Horses had greater use in terms of military strategy and load carrying, but even one cold winter or an acute summer drought could kill dozens of beasts, especially when deep snow covered the ground or when winter feed ran short. Mares were unable to suckle their young; starving animals would start dying a few months later, with the drought destroying not only breeding stock but a vital source of milk, cheese, and yogurt.[15] The nomads were forced to eat their dead horses. If the dry cycle lasted two or three years, the effects would be even more disastrous. Unable to find food, deprived of their horses, and incapable of defending themselves or raiding, they had no option but to join other groups, starve to death, or move. In some years, thousands of horses would perish. There was only one solution: move to better grazing. This usually lay to the south, on the margins of—or often on—lands settled by farmers.[16]

The gavotte between nomads and drought began long before Ginghis Khan and persists to this day. In it may lie one of the reasons why the Great Khan's hordes burst on an unprepared and unsuspecting world eight centuries ago.

THE STEPPES ARE a vast blank with respect to climatic study: instrument observations, even today, are few and far between. By the same token, historical records from medieval times are a precious rarity, and even those tell us little about climatic events. Russian climatologists have cataloged extreme climatic events such as major droughts in Eurasia since the early eleventh century, tracing cycles of exceptional warm and cold over thirty-year periods. They've linked these cycles with thirty-year temperature and rainfall records derived from local proxy data such as tree rings and hydrological information. The resulting temperature curves speak of a four-century warm cycle beginning about A.D. 850, with mild winters and dry summers, which coincides with warmer conditions in western Europe. Not that the climate was always benign. The *Chronicle of Novgorod* tells of catastrophic autumn rains in 1143 and 1145 that destroyed the harvest and caused hunger. The chroniclers also tell us that there were seventeen years of climate-induced famine during the early thirteenth century, culminating in a drought-induced famine in 1215 that forced the city's inhabitants to eat bark and sell their children into slavery. In 1230, another drought brought more suffering: "Some of the common people killed the living and ate them, others cut up dead flesh and corpses and ate them, others ate dogs and cats. . . . Some fed on moss, snails, pine bark, lime and elm leaves and whatever each could think of."[17] These disasters came at the height of well-documented warmer times in western Europe, when Norse ships still sailed to Iceland and Greenland and brought timber from Labrador. During the same period, the Slavs settled the coast of the Russian Arctic as far north as Novaya Zemlya, before the Little Ice Age descended.

The Novgorod chroniclers tell us that the climate of Eurasia's warmer centuries was never static; serious droughts and cold winters alternated with periods of quieter, more benign conditions, such as those of the early fourteenth century. Since the Ice Age, the North Atlantic Oscillation and its seesaw of atmospheric pressure between the Azores and Iceland have governed western European climate. (For the North Atlantic Oscillation, see sidebar.) High pressure over the Azores and low pressure over Iceland bring persistent westerly winds and mild winters. But when high pressure builds over Iceland and Scandinavia, winter temperatures plummet both in the west and on the steppes. Central Asia is remote from the moderating influences of the Atlantic and Pacific; continental weather systems bring dramatic changes in temperature and rainfall, altering the environment of the plains within days. Even a slightly late spring or a few weeks of summer drought can devastate a year's pasture. The records of the Novgorod chroniclers do not, of course, apply to the steppes,

The North Atlantic Oscillation

The North Atlantic Oscillation (NAO) is an irregular seesaw of changing atmospheric pressure between a persistent high over the Azores Islands in the Atlantic and a low that lingers over Iceland. The swings of the NAO are part of the complex atmospheric-ocean dynamic of the North Atlantic, which is still little understood. But the NAO is of critical importance, for it affects the position and strength of the North Atlantic storm track, which brings rain to Europe and parts of Eurasia. When low pressure persists over Iceland and high pressure builds off Portugal and the Azores, westerly winds persist over the North Atlantic, winter storms are strong, rainfall is plentiful in northern Europe, and winter temperatures are mild. Reverse the index from its "high" mode to "low," when pressure is high in the north and low in the south, and Europe suffers under much colder winter temperatures while the westerlies weaken. Bitterly cold air flows south and west from the North Pole and Siberia. No one has yet succeeded in predicting the swings of the NAO, which can endure in the "high" or "low" mode for seven years or more, even for decades, but is sometimes subject to rapid shifts.

Another pressure gradient also affects European winters. In extreme "low" modes, persistent high-pressure systems form between Greenland and Scandinavia. Temperatures then are higher than average in Greenland and much lower than normal in both northern Europe and eastern North America. When the pressure over Greenland is lower than in Europe, the temperatures are reversed, and European winters are milder. Such a "Greenland low" may have persisted during the warm centuries.

The behavior of the NAO depends on many complex factors, among them sea surface temperatures in the Atlantic, the mild waters of the Gulf Stream, and the powerful downwelling off southern Greenland that causes vast quantities of heavier, salt-laden water

from the Gulf Stream to sink far below the ocean's surface to fuel the ocean conveyor belt that circulates water through the world's seas. There are clearly links between the NAO and the complex gyrations of the Southern Oscillation in the Pacific (see chapter 9), which generates El Niños and La Niñas, but they are still poorly defined.

but we can be certain that the pattern of cycles of colder, wetter, and warmer and drier conditions applied over much of Eurasia.

DROUGHTS ON THE steppe are usually caused by persistent high-pressure systems over the Arctic. These systems, which can remain stationary for long periods, prevent the passage of the usual rain-bearing frontal systems and draw in intensely cold, dry air from northern seas. The fresh Arctic air enhances dry conditions. In 1972, for example, an anticyclone centered over Moscow persisted throughout the summer, blocking the passage of Atlantic depressions. Extremely hot, near-desertlike conditions in regions like the Volga and Ukraine cut summer rainfall to 20 or 30 percent of the average and resulted in very low relative humidity. Temperatures were 1.5 to 3.5 degrees F (3 to 7 degrees C) above the norm; the heat sucked such moisture as remained from the ground. Without doubt, similar intense droughts occurred in earlier centuries.

The medieval nomads were well aware of climatic variations from year to year. Long, snowy winters denuded pastures of their grass. Precious winter feed had to be stretched for an extra two months or more, being doled out in ever smaller quantities. Oxen and cattle subsisted off bedding grass and lost weight. Some became so weak that the herders had to help them rise. Calving losses rose sharply. Emaciated beasts perished of cold or were lost in the deep snow. In particularly cold winters, both animals and humans died in large numbers.

When the summer did come, it would arrive suddenly. The snow would melt rapidly, turning the plains into a quagmire, swelling streams, and impeding movements to summer grazing grounds. Soaring temperatures meant that little water percolated into the soil, grass growth was weak, and the summer graze was, at best, poor. The only protection against such disasters was movement. In the central regions of the steppe, the nomads traveled as far south as they could during the cold months to ensure that their grazing grounds were denuded of grass for the shortest possible time. In the summer, they moved far north to strategic, sheltered river valleys, areas where rainfall was slightly more plentiful and grass of better quality.

Water and its distribution across the landscape were also critical variables. Each tribe defined its territory around river systems, especially the sunken river valleys incised into the steppe that were the linear sinews of their territory. Nomads would winter in houses built at valley elevations below the level of the plain, then migrate northward in spring, sometimes as early as February or March in mild years, or as late as May in cold ones. The seasonal movement northward would stop and start, depending on local grazing and sometimes impeded by swollen streams. Eventually, their animals were allowed to graze on rich pasture, which might cover as much as 3,250 square miles (about 8,400 square kilometers). In mild years, people would sow grain, then neglect it until they were about to move south. In dry and cold years, they would have no chance to plant, for they would arrive at the summer grazing too late to sow, with too little time before cold weather killed the growing crop.

Each temperature change and rainfall shift dramatically altered the relationship between the nomads and their environment. Drier periods, with their life-threatening droughts, brought stunted pasture, decimated herds, extended searches for grass and water, and inevitable, often violent encroachment into neighbors' territories. In better-watered cycles, herds increased, the carrying capacity of the grazing land was much improved, and territories shrank, with a resulting decrease in warfare. For centuries, those living on the fringes of the steppe lived in fear of the fierce nomads, who would arrive without warning and create mayhem as they sought better pasture.

UNFORTUNATELY, THE EVIDENCE from climatic proxies that would enable us to identify the climatic fluctuations of Ginghis Khan's day is scant. Climate records extending back a thousand years are the Rosetta stones of paleoclimatology, rare chronicles treated with the scientific reverence they deserve, even if they are only proxies, indirect traces of ancient temperature and rainfall. Such sequences are rare treasures for archaeologists and historians seeking decadal and century-long climatic fluctuations like the Medieval Warm Period. As we have seen, the bolder among climatologists have combined these records into large-scale temperature reconstructions going back as much as a millennium. These portraits of past climate come, for the most part, from tree-ring records, historical documents, and instrument readings over the past century or more. But with respect to the climate on the Eurasian steppes at the time when Ginghis Khan embarked on his murderous campaigns, the record is still almost blank, except for broad generalizations and just one or two tree-ring sequences.

A research team from the Tree-Ring Laboratory at the Lamont-Doherty Earth Observatory in Palisades, New York, and the National University of Mongolia collected samples from living five-hundred-year-old Siberian pines at Solongotyn Davaa (otherwise known as Sol Dav), a location high in the Tarvagatay Mountains of west-central Mongolia.[18] The ecological conditions at this site are such that tree growth is influenced by temperature changes from one year to the next. After months of research, the team developed a temperature curve from living trees for the years 1465–1994. Then they returned for additional samples from well-preserved, long-dead wood, linking the tree rings from these fragments with those from the living pines. The expanded climatic sequence now extends back to A.D. 850 and, less reliably, as far back as A.D. 256, to the time when Rome was all-powerful and the Scythians flourished on the Eurasian steppes.

The Sol Dav leaves us in no doubt of today's warm temperatures, with the highest tree-growth rates coming between 1900 and 1999. But there are notable earlier periods of warmth around A.D. 800. The year 816 was the warmest in the entire sequence, even warmer than today, although 1999 was the warmest year over the past millennium. The warm cycle of the ninth century and another in the early 1400s bracket the centuries of the Medieval Warm Period. There was a period of colder conditions around 1100, so the warming was not monolithic. The cooling of the Little Ice Age ensued, five centuries of unpredictable cooling and warming that climaxed with very cold conditions during the nineteenth century.

Not only does the Sol Dav sequence provide evidence of medieval warming, but also its fluctuations coincide with temperature changes over at least the past four centuries recorded in the well-known Mann sequence for northern and western Europe already described in chapter 1. The well-preserved ancient pines of Mongolia place Ginghis Khan's conquests within an extended warm period during which frequent droughts may have wreaked havoc on steppe pastures in a world where people depended on horses as well as stock of all kinds. If the solitary Mongolian tree-ring sequence is a reliable barometer of the cyclical temperature and rainfall of the Great Khan's time—and there is every reason to believe that it is—then it's clear that the climatic pump of the steppes acted as it had for thousands of years, putting nomads and their restless movements into play on the steppe and bringing them into conflict with their neighbors to the south. The difference was that Ginghis Khan rose to power at a time when drier conditions shrank pasturage on the steppe. This was nothing new; but now, a brilliant leader succeeded in forging great armies of conquest from a maze of competing tribes and independent-minded chiefs. The Flail of God shook Asia and Europe to their foundations.

THE PROLONGED WARM period detected in the Mongolian tree rings coincides with Ginghis Khan's savage conquests: hotter and drier conditions would have meant a surge in warfare at a time of potential hunger and rising unrest. Ginghis Khan's incursions into China and his merciless smashing of the Seljuk Turks' Khwarezmid empire in central Asia in 1220 and 1221 brought the Mongols deep into settled lands.

Shortly before his death, in 1227, Ginghis Khan told his sons: "With Heaven's aid I have conquered for you a huge empire. But my life is too short to achieve the conquest of the world. That task is left for you."[19] The conquests continued after Ginghis was gone. One of his sons, Ögötai Khan, extended the empire westward in 1236. Batu, a grandson of Ginghis, soon conquered the Crimea, then ravaged what is now Bulgaria as well as fourteen Russian cities, turning their shattered remnants into vassal states. Next he turned his attention to Europe, with the objective of reaching "the ultimate sea." The Mongols under General Subutai divided into three groups, conquered Poland and Hungary, and swept into Austria, where they prepared for a probe into the heart of Europe in 1241. At that moment, Ögötai Khan died. Batu Khan was a potential candidate for Great Khan, so he withdrew his forces to the steppes. In the event, he was not chosen and devoted his efforts to consolidating his conquests around the Urals. He held sway over the Cuman steppes and over various Russian kingdoms and never returned to the scene of his former conquests.

BATU KHAN'S WITHDRAWAL coincided with the return of cooler, wetter conditions, which brought improved pasturage to the steppes. His kingdom flourished during generations of good pasturage, when warfare died down. Although Batu always maintained ambitions of returning west, good grazing conditions at home allowed his people to pasture a huge territory from the Volga–Don to Bulgaria. There were no incentives for ambitious conquests when grazing was plentiful and trade flourished with lands to the south.

But what would have happened if the climatic pendulum had not swung, and if droughts had intensified on the steppe? To judge from earlier centuries, warfare and restless movement would have continued, and, almost certainly, Batu Khan and his generals would have returned to the west. His spies had already given him a clear picture of the kingdoms that confronted them, and of their armies with their heavily armored knights, who had proved no match for Mongolian archers and horsemen. He would have followed his original plans, drawn up with General Subutai: invade Austria and destroy Vienna first, then move against the German principalities before turning his attention to Italy. If all went well, he would then have marched into France and Spain. Within a few years, perhaps as early as 1250, Europe would have become part of a huge western Mongolian empire.[20]

Would this have actually occurred? The Mongols had already defeated formidable European armies in decisive, bitterly fought battles on the Hungarian plain, where thousands had perished. That tales of their ruthless conquests and indiscriminate slaughter had preceded them would have given them a major psychological advantage in a Europe riven by factionalism and chronic rivalries. By the time Batu had mastered Europe, the Mongols would have accumulated a vast experience not only of conquest, but also of assimilating themselves with, and accommodating, other cultures and religions. If central Asian history is any guide, European civilization would have continued to flourish as the new conquerors became absorbed into its fabric.

But fascinating questions arise. Would Europe have become a Muslim continent, or would the Mongols, who were tolerant of other beliefs, have left the Catholic church alone? Had the conquest taken hold, would there have been incentives for European explorers and merchants to find new ways to tap the riches of Asia by opening new seaways across the Atlantic and round the Cape of Good Hope to India when they could have land routes across a unified empire? And what would have been the impact of the Mongols on Muslim Spain? Here one could expect the same process as happened in central Asia: an environment in which Islam flourished, and might even have expanded northward over the Pyrenees.

There would have been a point where the momentum of conquest slowed—perhaps when the conquerors reached the Atlantic, or perhaps earlier. Had the climatic pendulum not swung back, there would have been no incentive to return home to a drought-ridden, arid land. Nor would peace have descended on the steppes, where, under benign conditions and with ample pasturage, each summer the tribes moved from their winter quarters in the south near the Sea of Azov and Astrakhan-Serai to summer grazing on the banks of the Don and Oka rivers. For all the pull of the steppe and the nomadic life, the political and economic center of gravity of the Golden Horde's empire would have moved west into better-watered, more settled lands. But, equally well, as happened to Ginghis Khan's domains, the sheer size of the empire, corruption, and inefficient administration might have caused Europe to split up into a patchwork of states very different from those that witnessed the Renaissance and the Age of Discovery.

The ebb and flow of Mongol rule would have depended in part on the realities of nomad life, just as it had for thousands of years. When pasturage was good, there was peace; when the climate deteriorated and drought ravaged the steppes, warfare broke out and the people of the settled lands trembled in fear. The endless rhythms of warm and cold, plentiful rainfall and drought, ample grass and no forage were a major engine of history, as powerful in their way as economic changes, the ebb and flow of political intrigue, and the abilities of individual rulers. Ginghis Khan and his armies, and the smallest tribe on the great steppes, were affected by the same realities. When drought on the plains coincided with social unrest and brilliant generalship, the foundations of history shook. And had the droughts continued, European civilization might have a very different face today.

CHAPTER 4

The Golden Trade of the Moors

They start from a town called Sijilmassa . . . and travel in the desert as it were upon the sea, having guides to pilot them by the stars and rocks in the deserts.

—Anonymous, *Toffut-al-Alabi* (twelfth century)[1]

IN JULY 1324, THE SULTAN OF EGYPT welcomed a truly exotic visitor. Mansa Musa, ruler of the West African kingdom of Mali, was on a pilgrimage to Mecca. Hundreds of camels and slaves carried gold staffs and lavish gifts across the desert. Mansa Musa held court in Cairo for three months. To the Egyptians' astonishment, his subjects prostrated themselves before him and poured dust on their heads. The Malians injected so much gold into the Egyptian economy that the value of this most precious of metals decreased between 10 and 25 percent for some years. Tales of the African kingdom and its fabulous wealth reverberated through the Christian and Muslim worlds. By the end of the fourteenth century, two thirds of Europe's gold came from Mali, transported by camel across the Sahara. This "Golden Trade of the Moors" linked two very different worlds, those of the Mediterranean and of the western Sudan in West Africa, Bilad es-Soudan, what Islamic geographers called "the Land of the Blacks."

LIKE THE EURASIAN steppes, the Sahara Desert expands and contracts like a natural ecological pump. This, the southern margin of the medieval European world, is one of the hottest places on earth. Prevailing, dry northeasterly winds raise temperatures above 98 degrees F (37 degrees C) on more days of the year than anywhere else on earth. A dead environment, one might think, but the Sahara breathes, is never static. When rainfall increases by even a few millimeters, the desert's margins shrink, sometimes by many miles. In warmer periods in the remote past, vast shallow lakes and semiarid grasslands, watered by seasonal rivers from arid mountain ranges, covered thousands of square miles of what is now desert. Only one significant lake remains. One hundred and twenty thou-

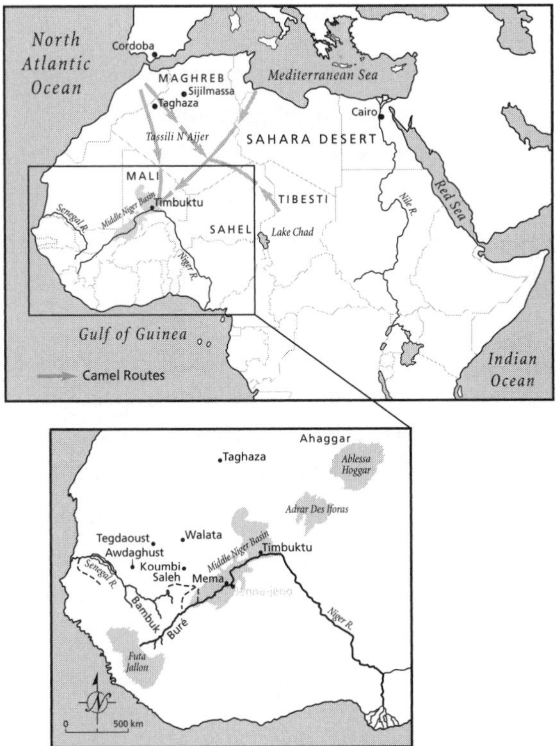

Locations mentioned in this chapter as well as a general impression of desert caravan routes.

sand years ago, Lake Chad, on the southern boundaries of the desert, covered a larger area than Eurasia's Caspian Sea. Today, Chad is drying up rapidly. During good rainfall years, the desert absorbs animals and people, often far north of Lake Chad. When drought years descend on the Sahara, water sources and grazing dry out, and the sparse desert populations move out to better-watered areas. The Saharan pump is never still; it is sometimes quiescent for a few years, then gyrates wildly during periods of highly variable rainfall from one year to the next. This is the story of the gold trade of a thousand years ago between the Islamic world and West Africa. The trade thrived through the warm centuries thanks to the highly adaptable camel and because those who handled the gold at the African end engineered their society to accommodate the sudden climatic extremes that marked the Medieval Warm Period.

THE CLIMATIC HISTORY of the Sahara and the Sahel, the semiarid grasslands that border the southern margins of the desert, is a relentless chronicle of chaotic shifts, well documented from both modern instrument records and proxy measures from deep-sea cores drilled off Mauretania.[2] We can even link some of these records to the deep-sea cores from the all-important Cariaco basin off Venezuela, described in chapter 8.

The Mauretanian sea cores reveal recent, abrupt changes of 1.0 degree F or more (2.16 degrees C) in the sea surface temperatures of the eastern North Atlantic. At the same time, changes in the salinity of the ocean at different levels can affect the workings of the ocean conveyor belt that is a fundamental driver of global climate, transferring heat as it does from the tropics to northern latitudes. The sea surface temperature in the eastern North Atlantic has a strong effect on the dry winds that blow across the Sahara. If sea surface temperatures are lower in the eastern Atlantic between 10 degrees north and 25 degrees north, and higher in the Gulf of Guinea, the monsoon winds are displaced southward, causing drought in the Sahel and Sahara. We know this because between A.D. 1300 and 1900, cooling documented in the Mauretanian sea cores caused dry conditions in the Sahel, including droughts that may have been worse than the disastrous megadrought of the 1960s. The cores allow us to make a tentative reconstruction of climatic conditions over the past two thousand years, and through the Medieval Warm Period, that goes something like this:

Between 300 B.C. and about A.D. 300, conditions in West Africa were stable and dry—as they were in both Southeast Asia and the Amazon basin—with rainfall somewhat below modern levels. People moved into better-watered areas like the middle Niger, where towns appeared.

After A.D. 300, rainfall increased to perhaps 125 percent to 150 percent of today's level until A.D. 700, a time when a formerly shrunken Lake Chad expanded dramatically. (There is no evidence of intervening dry periods, but they may yet be undetected.) Then, between A.D. 900 and 1100, there was an abrupt transition to much more unstable conditions, mirrored by increased monsoon variability in the Cariaco basin on the other side of the Atlantic. At some times there was high, stable rainfall; at others, drought. The Sahara's margins were constantly on the move.

In an attempt to understand these changes, the climatologist Sharon Nicholson has analyzed colonial meteorological records throughout tropical Africa and identified six different rainfall patterns, or climatic modes, through which African climate has cycled over and over again since the nineteenth century.[3] These modes alternate at random from extreme aridity at one end of the spectrum, characteristic of the Sahel in the 1890s and in the 1960s, through various intermediate and related stages, to disastrous wet shifts, when herds multiplied and overgrazed a new green landscape on the margins of the desert. Today, climate in the Sahel leaps abruptly and without warning from one mode to another in a completely unpredictable manner. It is likely that exactly the same kinds of abrupt shifts occurred during the Medieval Warm Period, creating extraordinary challenges for people engaged in cattle herding, subsistence agriculture, and long-distance trade.

Looking at these changes on a more global scale, we know that a dry year in the Sahel coincides with high pressure over the Azores and low pressure over Iceland. The northeast trades speed up and the Intertropical Convergence Zone stays well south.[4] Southwesterly winds bring less moisture to West Africa. When the sea surface temperature between 10

The Intertropical Convergence Zone (ITCZ)

The northeast and southeast trade winds meet near the equator, forming an area of low pressure. The winds converge and force moister air upward. As the air rises and cools, the water vapor condenses. A band of heavy rain forms, which moves seasonally toward areas where solar heating is most intense, places with the warmest surface temperatures. From September to February, the ITCZ moves toward the southern hemisphere, reversing direction for the northern summer. While the ITCZ moves over land, it shifts much less than over open water, maintaining a near-stationary position just north of the equator. Here rainfall intensifies as solar heating increases, diminishes as the sun moves away. As temperatures warm up, so rainfall increases, diminishing with cooling. El Niño events (see chapter 9 sidebar) have a major effect on the ITCZ, deflecting it toward unusually warm sea surface temperatures in the tropical Pacific and bringing less rainfall to the Atlantic and to the Sahara Desert's margins.

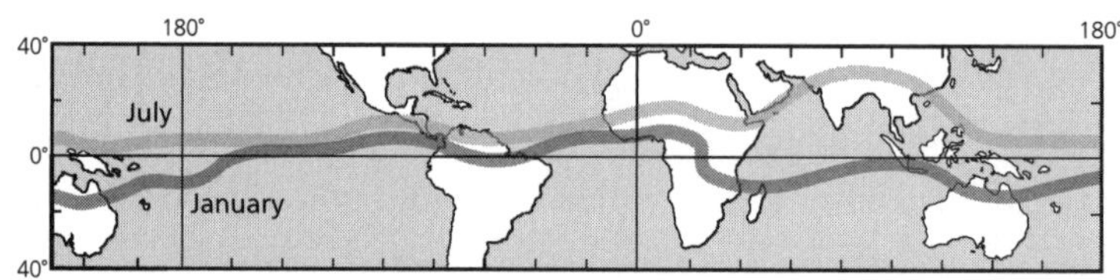

The Intertropical Convergence Zone and its range.

and 25 degrees north is 3.6 to 7.2 degrees F (2 to 4 degrees C) colder, and the Gulf of Guinea's waters are unusually warm, then the effect of the Intertropical Convergence Zone weakens. The deep-sea cores also show that the thresholds of many of the mode shifts are marked by very violent transitions in the interior, some of them preceded by extreme cooling spikes. One of these occurred in about A.D. 900, with another at the beginning of the eleventh century. Such unstable phases, with often prolonged droughts—and one stresses the word "prolonged"—would have been periods of remarkable difficulty and change for the Sahelians who experienced them.

WHAT EFFECTS, THEN, did the warmer temperatures and droughts of the Medieval Warm Period have on the Saharan gold trade and on the peoples of the Sahel? As far as the Saharan caravans were concerned, the effects were remarkably small because of the camel, or, more accurately, because of the saddles on camels' backs.

On the Eurasian steppes, life depended on cattle and horses, on good pastureland. When higher temperatures and drought descended on the grasslands, the nomads moved out in search of pastureland and water. The desolate plains and highlands of the Sahara were not a place where cattle and horses could thrive a thousand years ago, even if the rainfall increased slightly. In classical times, the desert was a fearful wilderness. Herodotus remarked that Libya on the Mediterranean coast was "infested by wild animals. Further inland from the part full of animals Libya is sandy desert, totally waterless, and completely uninhabited by anyone or anything."[5] Only a scattering of pastoral nomads survived near oases, and for them the margin between survival and starvation was always razor thin. Anyone who lived here was tough, resourceful, and constantly on the move.

The Romans turned North Africa into a prosperous granary, but never crossed the desert to the tropical lands to the south.[6] They lacked the pack animals that would allow them to travel for days at a time without water. To cross the Sahara with laden beasts on a regular basis meant combining highly adaptable behavior with an animal capable of going for up to ten days without water. That animal was the camel. And the camel was remarkably immune to severe drought.

The Golden Trade would never have thrived without the camel, but it was the development of a load-carrying saddle that made the camel the "ship of the desert." Camels store fat in their humps; their long necks allow them to browse in trees and brush; and their padded feet allow them to walk on soft sand. They conserve water through an efficient kidney system and they absorb heat by allowing their body temperature to rise significantly without perspiring. The Romans knew all about camels. They used them in North Africa to pull carts, and even as defensive barriers to protect soldiers.[7] They knew that these curmudgeonly beasts thrived in desert conditions. But these advantages were of limited value without an effective load-carrying saddle, which the Romans lacked.

The Saharan camel saddle came into use during the early Christian era, perhaps around the Nile Valley in what is now the modern Sudan, not for fighting, but for cargo. The saddle lies on the beast's shoulders forward of the hump, and so positioned it maximizes load-carrying capacity, endurance, and control. A Saharan camel driver steered his charge with a stick or with his toes. For the first time, camel caravans could now carry sufficient water and provisions (for the humans in the party) to cross long distances between oases from North Africa to the western Sudan.

No one knows when the first camel caravans traversed the western Sahara, but it was well before Islamic armies conquered North Africa in the seventh century. They followed obscure tracks that soon became well-established trade routes controlled by Muslim traders who came from a culture with a far broader outlook on the world than their predecessors from North Africa.

Saharan caravans followed a well-established routine. Heavily laden camels plodded southward from Sijilmassa each fall, south to Taghaza, where they picked up cake salt from nearby mines. Salt is a precious commodity for African farmers to this day, for they lack local supplies. From Taghaza they followed well-trodden paths to Walata, Ghana, and Jenne on the middle Niger River. The journey was hazardous under the most favorable circumstances. The desert was always hostile, even in times of slightly greater rainfall. Heat and dehydration were a constant threat. So were desert nomads, robed in blue burnooses, armed with gazelle-hide shields and spears, who would launch pitiless attacks without warning. Most caravan organizers negotiated agreements with nomad chiefs allowing them safe passage through the oases they controlled. The nomads also provided guides, who used rocky outcrops and the stars to navigate. They also provided camels to the merchants, who sold them back at the end of the journey.

The caravans were well-organized convoys. The camels laden with merchandise were supported by numerous others carrying water and provisions, or serving as mounts. Safety came in numbers—safety from raiding nomads, in a larger number of beasts who carried water and food, in the ability to transport larger loads and to make a great profit. During the twelfth century, some caravans numbered as many as twelve hundred to two thousand beasts. The journey itself lasted between six weeks and two months, with departure in the autumn. Writes the contemporary Muslim geographer al-Idrisi: "The camels are loaded at a very early hour and one travels until the moment when the sun appears on the horizon and the heat generated on the earth is unsupportable."[8] The caravans would rest until late afternoon, then proceed silently through the night, guided by the stars, just as they still do today.

Camel caravans made the long journey across the Sahara even during the driest years of the Medieval Warm Period. Those who traversed the desert spent a great deal of time acquiring intelligence about water supplies, for wells and oases were vital to a safe journey. Conditions were never the same from one year to the next. The cycles of wetter and drier conditions affected the patterns of the trade. When conditions were wetter, large numbers of wells were dug in the aqueous gravels of the central Sahara, around the highlands of the Ahaggar and Adrar des Iforas. Many caravans then followed direct routes over the dunes of the central Sahara to Taghaza and the town of Awdaghust (in modern-day Mauretania) on the borders of the desert, an important salt-trading center. During dry cycles, the caravans would follow more roundabout routes far to the west, or, passing eastward and northward from the Bilad es-Soudan, they traveled to the Adrar des Iforas, then west, ending up ultimately in Sijilmassa. The versatility of the camel provided sufficient flexibility to ride the gyrations of the desert pump.[9] The numbers of dead and exhausted beasts could be enormous; casualties were often in the hundreds from one caravan alone. The bleached skeletons of camels and their drivers littered the routes, but the Golden Trade never ceased. The camel and its load-carrying saddle proved an effective weapon against heat and drought even in the worst years, when extreme aridity affected cattle people living far south of the desert.

We know enough about climate change during the warm centuries to be virtually certain that they were a time of abrupt and sudden rainfall shifts. The Saharan pump would have moved into frenzied activity as the desert margins advanced and retreated even on a yearly basis. The Islamic discovery of the West African gold trade seems to have coincided with the end of a period of relatively stable conditions, with at least some more rainfall than today. Water holes would have been more plentiful, and desert travel by camels, while yet perilous, could be organized on a relatively large scale. Fortunately for the outside world, the adaptability of the camel and the skill of those who lived in the desert and on its edges gave the Saharan gold trade a considerable degree of immunity from climatic shifts.

The human bridge was as important as the camel. Much of the trade depended on the nomadic Berbers, ancient inhabitants of the desert, who bred camels and also accompanied many of the caravans. They lived at both ends of the trade routes and served as the human link between north and south. The other connection was Islam, which was eventually to become the common religion of the North African merchants, Saharan nomads, and many African rulers and traders south of the desert.

Gold was of profound importance in the Muslim world, which was a strong incentive for the trade to overcome the hazards of desert travel. The golden dinars minted by the caliph in Baghdad, and by the caliph alone, circulated throughout the Maghreb (northwestern Africa) and Spain. At first, gold supplies came from booty taken in Syria and Egypt, also from Christian treasuries and sources in upper Egypt and farther up the Nile. But by the eighth century, West African gold was already well known. The metal itself arrived as dust, traded from miners in the

Bambuk region of the Senegal River, twenty days' journey south of the kingdom of Ghana in the Sahel, then a major staging post for the gold trade. Enterprising merchants tried to gain control of the gold sources, but to no avail. The miners firmly maintained their independence and little was known about their operations. They extracted the ore from auriferous river gravels by digging numerous small pits. But the yield from these simple workings was enormous. The Baghdad astronomer al-Fazari, writing late in the eighth century, called Ghana the "land of gold."

In A.D. 804, the rulers of the Maghreb began using Sudanese gold to mint their own dinars. Sudanese gold financed wars of conquest and brought immense wealth to Islam. Until the twelfth century, most West African gold remained in the Muslim world. Western Europe had abandoned gold-based currency, partly because an adverse balance of trade with the east had drained its supplies with few means of replenishment. As Europe's economies recovered and the Italian cities built powerful fleets to combat Arab piracy, a growing volume of trade in cloth and other commodities attracted increasing quantities of gold. By the end of the thirteenth century, European mints were making gold coins. Country after country returned to the gold standard. The demand for gold increased; prices rose, then stabilized. Most of late-fourteenth-century Europe's gold came from the western Sudan. The relative immunity of the camel to the ravages of the desert pump ensured that the trade continued to help change history.

No one knows exactly how much gold passed into the trans-Saharan trade from West Africa. Tax records levied on caravans at Sijilmassa during the tenth century and quoted by the author Ibn Haukal cover imports of some 9.4 tons (8.5 metric tons) of gold annually, perhaps half of an annual total of some 16.5 to 18.7 tons (15 to 17 metric tons), carried northward from West Africa. In 951, Ibn Haukal saw a promissory note for 42,000 dinars drawn on a merchant in the north, a measure of the staggering wealth of the trade in its heyday.[10]

Where, then, did the gold come from? Before traveling farther south, the caravans stopped at Awdaghust at desert's edge, a large and populous Berber town of flat-roofed mudbrick and stone houses overlooked by a high outcrop. In Awdaghust's always crowded market, one could buy salt, sheep, honey from the Sahel, and food of all kinds—provided one paid in gold. The prosperous oasis town had good water and was home to merchants with a monopoly over the trans-Saharan trade. They organized their caravans under the auspices of the Sanhaja nomads of the desert. The Muslim geographer al-Bakri tells us that the nomadic Sanhaja ruler of the town's domains extended over a distance of two months' traveling. He was said to be able to field one hundred thousand camels. Gold and salt flowed through the town, whose leaders were careful to maintain good relations with powerful chiefs to the south, especially those presiding over a gold-rich kingdom named Ghana.

> The King adorns himself like a woman wearing necklaces round his neck and bracelets on his forearms and he puts on a high cap decorated with gold and wrapped in a turban of fine cotton. He holds an audience in a domed pavilion around which stand ten horses covered with gold-embroidered materials . . . and on his right, are the sons of the vassal kings of his country, wearing splendid garments and their hair plaited with gold.
>
> At the door of the pavilion are dogs of excellent pedigree. Round their necks they wear collars of gold and silver, studded with a number of balls of the same metals.[11]

Al-Bakri's description of Ghana was the stuff of legend. He never visited the Sahel, but drew his account from sources in the Córdoba archives. His Ghana was a Mediterranean-style court deep in Africa, a capital with two towns, one with twelve mosques where Muslim merchants dwelled, the other the ruler's compound, with sacred groves and royal tombs, nearly ten kilometers away. The royal treasure included a gold ingot said to weigh nearly 30 pounds (13.6 kilograms), so large that it became famous through the Christian and Muslim worlds.[12]

This seemingly imposing capital, Koumbi Saleh, is commonly thought to have lain about 300 miles (480 kilometers) west-southwest of Timbuktu and the Niger River bend. There are indeed extensive stone ruins here, also Arabic inscriptions, but no traces of the royal quarter nor of the burial mounds said by Muslim travelers to lie close by. The ruins lie at the extreme northern limits of the Sahel, where agriculture would have been near impossible even in periods of higher rainfall.[13] Maybe Koumbi was not Ghana's capital at all, but a small trading community, part of an entirely different, more decentralized kingdom. For the moment, the kingdom of Ghana remains elusive, its capital peripatetic. Our only certainty is that it was not an Islamic polity, but an indigenous African domain, something very different from al-Bakri's portrait, with roots deeper in West Africa, where the gold came from.

For a long time, the gold sources were a mystery. Writing in A.D. 872, the historian al-Yaqubi repeated an oft-told tale about gold sprouting from the ground like carrots. As always happens with gold, the fables grew with the telling, until they produced an "Island of Gold" where gold was to be had for the taking.[14] The miners were well aware of gold's value and kept the locations of their ore deposits a close secret, lest outsiders try to take control of the supply. For this reason, they refused to trade face to face: the merchants piled their goods, mostly cake salt, on the riverbank and passed out of sight while the local people placed heaps of gold alongside each pile. If the visitors were satisfied, they would take the gold and retreat, beating drums to signify the end of the transaction. On one occasion, they captured a miner in an attempt to find the source of the gold. He pined to death without revealing anything. The trade ceased altogether for three years before resuming.

The miners of Bambuk and Buré, another area to the east, were timid, private people who jealously guarded their gold-mining activities, which is why they engaged in silent trade. No Berber merchant from the Sahara ever visited the goldfields, so the Island of Gold remained a mystery. It remains a geographical conundrum today. During the twelfth century, al-Idrisi described the island as an area nearly 300 miles long and 155 miles wide (500 by 250 kilometers) that flooded each year, where the local people "collected gold." The position of the island on his map coincides with that of the seasonally flooded middle Niger delta, inhabited by Mande-speaking farmers and fisherfolk.

THE NIGER IS one of Africa's great rivers, rising in the mountains of Guinea near the Sierra Leone border, then flowing northeastward into a great inland delta, a patchwork of tributaries, channels, swamps and lakes. This interior floodplain is what the archaeologist Roderick McIntosh calls "a vast alluvial garden abutting the bleak Sahara."[15] Here, the desert caravan networks came in contact with much older riverine trade routes. The middle Niger floodplain was rich in grain and other basic commodities, including potting clay, but, like Mesopotamia, lacked stone, metal ores, and salt. Over many centuries the Mande farmers and fishers of the region developed a lattice of contacts with other peoples near and far to supply their needs. They were also active players in the Saharan gold trade.

The Mande (the term refers to a language common to many groups) are descended from Saharan peoples, and settled in the Sahel during a series of dry spells that affected the region about two thousand years ago, perhaps earlier. Today, Mande speakers flourish over a large area of West Africa from Gambia to Côte d'Ivoire. They were millet farmers and cattle herders, also traders, who exchanged copper, salt, and semiprecious stones with contacts in the desert. As they moved southward, many of them colonized the fertile basins of the Niger River.

Today, the annual flood inundates some 21,000 square miles (55,000 square kilometers) of swamps and lakes, but covered a much larger area at wetter times in the past.[16] The floodplain environment is diverse, unpredictable, and made up of radically different landforms and soils packed closely together. The Mande who live there are the Bozo, who are fisherfolk, and the Marka, who cultivate many varieties of African rice (*Oryza glaberrima*). The Bozo are constantly on the move, their lives dictated by the breeding cycles of both small fish and the enormous Nile perch. Sometimes as many as 150 canoes will gather around artificial barrages where they harvest enormous catches. The Marka help with the major fish runs. In return, the Bozo assist the Marka, for the rice harvest comes during high flood, when fishing is poor.

The Marka, farmers, traders, artists, and musicians, farm in a remarkably stressful environment. A sudden flood and torrential rains can wipe out a year's hard work in a week. Early or late floods can leave villages desolate. Irregular rainfall at the beginning of the rainy season, or arid years, play havoc with newly prepared fields, quite apart from the depredations of rice-eating birds and rodents. The Marka combat these uncertainties by growing several varieties of rice. But, above all, their success depends on a reservoir of weather-forecasting knowledge acquired over many generations. Whereas the unique qualities of the camel gave caravan leaders the ability to ride climatic punches, the people of the Niger basin adapted their society to their hostile and unpredictable environment with a brilliant combination of social engineering and ritual observance.

The societies that flourished in the middle Niger a thousand years ago thrived on constant change in a place where different cultures lived in an exceptionally heterogeneous environment. They thrived not only by cultivating a wide variety of crops, but also by making extensive use of social memory. Roderick McIntosh calls the middle Niger basin a "symbolic reservoir," a place where a shared body of social values that originated in deep time survived over thousands of years to define history and society. This was not a world where highly centralized, authoritarian kingdoms flourished, with all power flowing to the center, as once was the case in ancient Mesopotamia with its competing city-states. There was no hierarchy of power, as there was among the ancient Maya or in medieval Europe. Here, powerful kin groups and people engaged in activities of all kinds, living together under a system of unspoken checks and balances that gave everyone considerable autonomy, mutual support, and enhanced chances of survival in the unpredictable desert-margin climate.

One of Africa's longest-inhabited archaeological sites provides a telling portrait of the developing Mande world. The ancient mound known as Jenne-jeno lies 1.8 miles (3 kilometers) south of the modern town of Jenne in the upper Niger basin.[17] The site lies at a strategic location near basins where one could grow climatically tolerant *Oryza glaberrima*, close to pastureland and open plains, with access by boat to the Niger. Jenne-jeno probably came into being as the drying Sahara expelled people to its margins after 300 B.C. People continued to live there for over sixteen hundred years, at a location that remained dry even in the highest flood years. Few places anywhere, let alone in West Africa, have such a long history. The settlement expanded vertically and horizontally, from about 49 acres (20 hectares) in A.D. 300 to nearly double that size three centuries later. Jenne-jeno's people lived off agriculture, hunting, plant gathering, and fishing; this generalized subsistence economy changed little over the centuries despite major population increases and rapidly changing climatic conditions. Through the town's long history, its inhabitants resolutely maintained a highly diverse economy, exploiting numerous microenvironments rather than seeking to increase food supplies with irrigation works or raised fields, as was the case with the Maya, for example.

At one point in Jenne-jeno's later history, no fewer than sixty-nine settlements lay within 2.4 miles (4 kilometers) of the main settlement. But why did the people here and elsewhere choose to live in clusters of villages instead of in densely inhabited cities like those of the Islamic world? The reason was climatic. The Mande lived with a constant backdrop of sudden, potentially disastrous climate change and they engineered their society around this reality.

For centuries, so excavations tell us, the Jenne-jeno people maintained a lifeway that involved agriculture, fishing, gathering, and moving within the local environment when necessary. But they were far from being stationary pawns affected by short-term climatic cycles. McIntosh believes the Mande were far more proactive, combating unpredictability by linking together farmers and specialized artisans such as metalsmiths into a generalized economy. People lived at separate, clustered locations, linked by ties of kin as well as by powerful myths and legends that provided rationales for decision making. No charismatic leaders, no cities or powerful elites, no armies to enforce law and order held this system together. Instead a "weather machine" of belief and ritual provided a framework for predicting rainfall and drought.[18] Social memory, carefully preserved by select people, is the core of the Mande weather machine.

How can we know how ancient communities responded to climate change, to perceived changes in their environment? We cannot reconstruct ancient minds. But we can examine Mande social memory, which couples their existence to the real world in complex ways. People surely had social memory of climate change, of catastrophic droughts and floods, perhaps, like today, associated in their minds with the names of individuals who were victims of the disaster and even named after them, or with a group of people, such as ironworkers, who are perceived to have occult powers. They preserve generations of knowledge about climatic shifts and environmental conditions, and often predict impending changes and offer strategies to combat them. At issue here is the question of authority to make decisions for the future, where climatic and other hazards wait unseen. Who can handle such matters? Who can be trusted not to abuse the dangerous knowledge of climate change and appropriate social responses for personal gain? In Mande society, climatic prediction is not the disinterested, scientific forecast of a climatologist. Every predictor also has responsibilities in the domain of social action, so his or her predictions are a critical link between climate change in the objective world and the perceived world on which people are going to act.

The dynamics behind the Mande weather machine are based on a set of familiar cultural values enshrined in many layers of legends and symbols. Numerous interest groups make up Mande society, all negotiating with one another constantly over land and the events of history, as they have done for centuries. The result is a powerful landscape of what one might call social memory, defined both by oral traditions and by recollections of climatic and other events. The fluid adaptability of their society has always depended on this familiar and ever-changing social landscape. The Mande had to be watchful and flexible in their responses to the very complex, variable, and unpredictable Sahel climate. Violent shifts from ample rainfall to drought, or modes in between, were part of the ecological and social crises that shaped core values and transformed authority. This is why the people of Jenne-jeno and other Mande towns lived in clusters, in a heterarchical society that created flexible and changing social landscapes more responsive to climate change than highly centralized contemporary societies such as the ancient Maya or the Chimu of northern coastal Peru.

The men and women who exercised the greatest influence in Mande society were members of secret societies. The *komo* remains a pervasive form of secret society, led by the *komotigi*, today often a blacksmith. The original secret societies were for hunters, long before metallurgy arrived in the Niger basin. A *komotigi* had the ability to see into the future; he was a curer and a protector against malevolent spells. An astrologer and weather forecaster, he studied the heavens and the most visible celestial bodies. A *komotigi* was expert in the behavior of animals and plants, which he used to predict whether rain would come when needed at planting time. The *komo* persists, as do extremely secretive hunters' societies. Today, there are at least seven major secret societies in the middle Niger.

The earliest specialists in Mande country were hunters with occult powers and social authority. From very ancient times, such individuals traveled to special places that were imbued with spirits, and as such were powerful and dangerous. Here they harvested power, the ability to control weather and other aspects of life. Over two thousand years ago, with the coming of agriculture and ironworking, metalsmiths became important *nyama* holders. (*Nyama* can be loosely defined as the earth's force.) The smiths were members of the first secret societies, the central actors in centuries of Mande resistance to hierarchy and centralization, armed with a symbolic and mythic repertoire that confronted environmental and social stresses. For centuries, Mande secret societies dispersed knowledge about the landscape over enormous areas.

The great Mande folk heroes harvested the power of *nyama* to navigate safely through a dangerous world. For instance, the smith-sorcerer

Fakoli undertook quests to acquire the medicine items such as birds' and snakes' heads that hang from his sorcerer's bonnet. Fakoli's quest was also a knowledge journey, which took him on an expedition through a spiritual and symbolic landscape. Another mythic hero, Fanta Maa, a Bozo culture hero, learned how to become a proficient hunter while still in his infancy. Under threat of extinction, the animals came together and selected a gazelle to take the form of a young woman and seduce Fanta Maa, then lure him into the wilderness and to his death. But Fanta Maa used hunters' paraphernalia to unravel the plot.

Nyama is malign, if controlled, energy that flows through all animate and inanimate beings. As the Mande perceived climate change, it was because of perturbations in *nyama* within the landscape. Generations of understanding how *nyama* perturbations affect the environment allowed Mande heroes and *komotigi* to predict climate change. The most powerful among them had the spiritual authority to manipulate the forces that bring on rains or drought—or to kill someone at a distance with merely a thought. Thus, charismatic individuals moved through a symbolic landscape in which they harvested power, authority, and knowledge. Hunters moved into the wilderness on tours of spiritually powerful locations. There they killed spirit animals. The killing released vast amounts of *nyama* that only the best hunters could control, and ensured well-watered lands for stressed communities.

Hunting lodges are still revered places where the accumulated climatic lore of generations, about resources within local areas, resides. Even famous hunters still go to such lodges to acquire more and increasingly obscure knowledge. Underground watercourses may also mark former north–south migration routes used in better-watered times. To Mande farmers, the landscape was, and still is, a catalog of names and places, a predictor of human occupations and climatic shifts that was very successful in combating drought and flood over many centuries.

The Mande weather machine worked well. By A.D. 800 to 900, Jenne-jeno was 0.6 mile (1 kilometer) long, with, by conservative estimate, as many as 27,000 people living in the town and in the sixty-nine satellite communities within a 2.5-mile (4-kilometer) radius. Between A.D. 300

A Mande hunter with his snake avatar.

and 700, local rainfall was about 20 percent higher than that between 1930 and 1960. After 1000, the climate was much more volatile and the town went into decline. The people abandoned low-lying rice-growing regions, then higher and lighter floodplain soils, and switched from rice to drought-resistant millet, displacements caused in part by changes in the river flood regime, at a time when drier conditions were setting in. In nearby Méma, large, clustered mounds near channels and water-filled depressions give way to smaller, more isolated settlements, often on sand dunes. Everywhere in the middle Niger, communities large and small adapted to new circumstances, not necessarily without suffering, but just as their ancestors had.

THIS, THEN, WAS the cultural environment from which Ghana emerged around A.D. 700, at the end of a period of relative climatic stability.

Mande oral traditions tell of a great hero, Dinga, grandfather of the Ghanaian chiefs, a brilliant hunter, an extractor of *nyama*. Ancient tales describe his movements across the landscape, the power places where he triumphed over guardian and knowledge animals. Dinga settled at Jenne for twenty-seven years and took a wife, but had no children. Perhaps this was a time of forging solid alliances with Jenne, which was not under Ghanaian rule, but always friendly to its neighbor. Dinga moved on and settled to the northwest, vanquished a female protector spirit at Dalangoumbé, and patched together the kingdom of Ghana, probably through a long process of alliance building, of constructing agglomerations of social groups. The same thing happened again and again in the Sahel, the creation of loosely consolidated groups of chiefdoms that paid tribute to the core. Thus it was that the chief of Ghana was able to tax the trade in gold and other commodities from the south. At the same time, he maintained a standing cavalry to deal with hostile camel nomads and uncooperative chiefs.

Geographer al-Bakri's descriptions of Ghana appear during a few centuries of stable and relatively favorable climate. The Ghana he described was a fluid kingdom, marked by its heterogeneous organization and by its flexibility, forged partly by conquest but more importantly by the constant checks and balances that had always been a part of Mande existence. Ghana indeed possessed great wealth, but its greatest riches came not from gold and material things, but from the rich practices and traditions of its native culture that enabled its members to thrive in a climate of pitiless and violent extremes.

At the end of his life, the traditions tell us, Dinga passed on his accumulated *nyama* to his son, the water snake Bida, twin of the Ghanaian founder chief Diabé Sissé. Bida agreed to provide sufficient rain and gold from Bambuk, several days to the southwest, if he was given the kingdom's most beautiful virgin each year. One year, the virgin's suitor killed the snake. The lopped-off head bounced seven times and landed in gold-rich Buré, much closer to Mali. Gold production moved away from Bambuk and toward Mali. Seven years of drought and famine devastated Ghana. The kingdom fell apart. Once the spirit animal died, malevolent *nyama* intervened. It was no coincidence that Ghana ran into problems with its desert neighbors just as the period of relative climatic stability ended.

Oral traditions are history filtered through fickle human memory. Perhaps the tale of Bida's demise and the droughts that followed constitute a dim recollection of violent climatic swings, perhaps an epochal drought that caused an always flexible kingdom to evaporate. At a time of rising Muslim influence and conquest, Ghana remained pagan until 1076, when the Almoravid chieftain Abu Bakr captured Koumbi and imposed Islam on its inhabitants.[19] A century and a half later, in the east, the kingdom of Mali rose to prominence, the golden empire made world famous by Mansa Musa's hajj in 1324.

CHAPTER 5

Inuit and Qadlunaat

They made their ship ready and put out to sea. The first landfall they made was the country that Bjarni had sighted last. They sailed right up to the shore and cast anchor, then lowered a boat and landed. There was no grass to be seen, and the hinterland was covered with great glaciers, and between glaciers and shore the land was like one great slab of rock. It seemed to them a worthless country.

—*Graenlendinga Saga* (twelfth century)[1]

A.D. 1000. WISPS OF FOG HOVER OVER the wave tops of the Davis Strait west of Greenland in a delicate tracery against the surrounding gloom. The bearded Norse skipper peers into the gray, oblivious to the pervasive chill. His crew huddles in their thick cloaks, except for the helmsman manning the great steering oar at the stern. Two young men fill the time by sharpening their iron swords, rubbing fat over the shiny surfaces to prevent rust. The square sail creaks and groans as the boat rises and falls in the ocean swell, the flexible hull twisting effortlessly with the waves. The ship sails slowly, a tiny island in a featureless wilderness of heaving waves and murk. A chill north wind whispers through the gloom, just giving steerageway, nothing more. The young crewmen have been through uncertain situations before—endless waiting hove to in a gale, days of drifting aimlessly in a calm far from land.

The hours pass slowly as the fog thickens, lightens momentarily, then returns. At last, the wind shifts to the east and strengthens. The light air becomes a good sailing breeze. The fog quickly dissipates to reveal a clear, hard horizon and a deep blue sea. The steersman shouts and points ahead. Jagged, snow-covered mountains stand boldly against the now bright sky, highlighted by the late afternoon sun. A collective sigh of relief passes through the ship. If the wind from astern holds, they will reach a sheltered anchorage among the islands to the west the next day.

When the Norsemen reach land, they know they will encounter Inuit hunters, who subsist as they always have off fish and sea mammals.[2] They have come in search of walrus ivory, but have only one thing in common with the indigenous inhabitants—iron.

Eurasia and the Sahara may have suffered from drought, but the warm centuries were a boon to those in the Arctic. They brought less severe ice conditions in the far north, and saw a surge in Norse voyaging to the west, to Iceland and beyond. This is the story of how, during the warm centuries, favorable ice conditions in the North Atlantic and in the Canadian Arctic brought two completely different worlds into transitory contact—those of the Norse and of Inuit peoples whose ancestry extended as far west as the Bering Strait.[3]

As in Europe, the Medieval Warm Period brought milder winters and a longer growing season for cereals in much of Scandinavia. Population densities rose, creating land shortages and limiting opportunities for young men. They lived in a volatile society, riven with quarreling, factionalism, and violence. Each summer, young "rowmen" left in their long ships in search of plunder, trade, and adventure. As ice conditions in the north improved and the Arctic pack receded, Norse skippers, long expert coastal navigators, ventured farther offshore, into the North Atlantic.

Contrary to popular belief, the Norse never sailed far offshore in

Locations mentioned in this chapter.

their warships. Instead, they relied on the *knarr*, a stout, load-carrying merchant ship that was capable of sailing far across the ocean. *Knarr*s were light but strong, easily repaired at sea or on remote beaches. The crew subsisted off stockfish, the dried cod cured in the chill Arctic winds of spring in the Lofoten Islands off northern Norway. As the warming began, experienced skippers sailed west into waters unknown except to a handful of Irish monks who had traveled as far as Iceland in large hide boats a few generations earlier. Few of the adventurous crews wrote of their experiences, which passed into the legends commemorated by the Norse sagas. Many ships never returned, wrecked on fretted, stormbound shores or foundering without trace in savage offshore gales. But Norse colonists settled in the Orkneys and Shetlands off northern Scotland soon after 800 and on the Faeroes shortly thereafter. In 874, the Norseman Ingólf landed on Iceland. By 900, colonists had settled on the island, bringing their dairy economy with them. At the time, the winters were milder than they had been for centuries. Today, the southern boundaries of the northern pack ice lie about 60 miles (100 kilometers) off Iceland's north coast; when the first Norse settlers arrived, the ice was at least twice that distance offshore. Even in these milder conditions, life in Iceland was harsh, especially after a cold winter. The colonists combined dairying with seal hunting and inshore cod fishing. Milder summers allowed them to grow hay as winter fodder and to plant barley until the twelfth century, when cooler conditions made cereal cultivation impossible again until the early 1900s.[4]

In about 985, Eirik the Red, exiled from Iceland after some killings resulting from family feuds, sailed west and settled in southern Greenland. There he found better grazing than at home. Soon two colonies flourished, one in the sheltered waters of Greenland's southwestern coast, another farther north in the modern-day Godthåb district, at the head of Ameralik Fjord. The settlers found themselves on a coast that was ice free for most of summer at the time, warmed by the north-flowing Greenland current that hugs the shore. This favorable current carried the colonists' fishing boats into the fjords and islands around Disko Bay, to a place that abounded with cod, seals, narwhals, and walrus. Here, in what they called Nordrsetur, they collected enough ivory to pay tithes to diocesan authorities in distant Norway for many years.[5]

The West Greenland current flows into the heart of Nordrsetur and Baffin Bay, where it gives way to much colder, south-flowing currents. Even a modest sail offshore would have brought the Norse in sight of the snow-clad mountains of Baffin Island on the other side of the Davis Strait, which is little more than 200 miles (325 kilometers) wide at its narrowest point. The colder waters on the west side of the strait along Baffin Island, Labrador, and eastern Newfoundland experience a longer ice season and heavier ice cover that can last well into summer. But during the milder centuries of the Medieval Warm Period, when the ice cover dispersed relatively early in the season, coasting along eastern shores may have been considerably easier and less hazardous.

We do not know when the first Norse ships landed on Baffin Island, but it may have been before the first documented sighting of Labrador by Bjarni Herjólfsson in about A.D. 985. Lost in fog and light north winds on a passage from Iceland to Greenland, he eventually sighted a low-lying, forested coast quite unlike the glaciated mountains of his intended destination. He sailed back without landing and was criticized for it. Then followed the famous voyage of Leif Eirikson, the son of Eirik the Red, who anchored off a rocky, icebound coast, then coasted southward before a northeasterly wind past "Markland," the southern Labrador coast, until he reached the mouth of the St. Lawrence River, then southward to a region south of the great estuary that he named Vinland, on account of the wild grapes that grew there. He founded a small settlement at L'Anse aux Meadows on the northern peninsula of what is now Newfoundland. The settlement remained in use for a few years.[6]

Later expeditions in search of Labrador's timber came in contact with numerous indigenous Beothuk people, who fought them so fiercely that the Norse never settled permanently on the western shore. "When they clashed there was a fierce battle and a hail of missiles came flying over," the *Vinland Sagas* tell us.[7] For two centuries, Greenland ships sailed north and west, then used favorable currents to coast southward. Once they had built ships, or simply acquired timber, they would sail directly home on the wings of the prevailing southwesterly winds. Natives and newcomers seem to have avoided one another.

Sailing the eastern shores of the Davis Strait and down Labrador was fraught with danger, even in the warmer centuries. The crews faced hostile native Americans, polar bears, icebergs, and sudden storms offshore in notoriously windy waters. Navigation in icy waters and close inshore was far more dangerous for the Norse in their wooden boats than for the Inuit, who relied on light kayaks and hide boats that were easily pulled out of the water while being relatively immune to puncturing and easy to repair. A sudden freeze could crush a *knarr* in minutes, even in summer.

As much as they could, the Norse steered well clear of the margins of the pack and breaking summer ice. But, for all the dangers, abundant cod and milder conditions allowed the Greenlanders to voyage freely across the Davis Strait and into narrow channels of the Canadian Archipelago. There they encountered Inuit hunters, who welcomed them, for the Inuit craved what to them was an exotic substance: iron. Unlike the insular Beothuk of Labrador, the Inuit were part of a far larger Arctic world, linked by informal trade networks to other hunting groups with common ancestries that extended as far west as the Bering Strait.

Can we, then, trace this trade westward and connect it to the warm centuries?

A.D. 1000. THE Bering Strait is a sullen gray wilderness of ice-strewn waves. The wind is calm, the temperature near freezing, and the surface of the water unruffled. Low clouds hover. The hunter sits absolutely still in his skin kayak, his eyes quietly scanning the dark ocean. His hunting gear lies close to hand, his paddle barely touching the water. Life on the water is second nature to him, often more comfortable than being on land.

A black head surfaces momentarily, close downwind. A seal looks around inquisitively. The hunter waits, the kayak motionless. His prey slips under the water, leaving only ripples behind. Now the familiar vigil begins, the boat dead in the water. As he waits, the hunter checks his harpoon and the coil of line attached to the razor-sharp head. The waiting stretches from morning to afternoon. He sees the seal once more some distance away; again it vanishes. He paddles softly toward the dispersing ripples, and then waits again. Suddenly, the prey surfaces within harpoon range. The Eskimo casts his harpoon. The iron tip sinks deep into the seal, which dives at once. The line with its floats courses out as the shaft detaches from the head. For a couple of hours, the hunter follows the bobbing float as his quarry weakens, then dies. When the carcass comes to the surface, he lashes it to his kayak and heads for home.

The Bering Strait is a harsh, unforgiving place where winters can last nine months a year. Dense fogs mantle the gray ocean for days on end, cutting visibility to a few yards. Howling winds blow through the mists. Except during the brief summer, fractured pack ice crowds the narrows, occasionally surging ashore in violent storms. The Siberian coastline is the more rugged side, with steep cliffs and clear landmarks. Low-lying coastal plains, numerous lakes, and low, rolling hills mark the Alaskan side. The best sea mammal hunting lies on the western shores, protected by strategic promontories. Twelve hundred years ago, as the Norse *knarr*s were crossing the North Atlantic for the first time, elaborate sea mammal and caribou hunting societies thrived on both the Asian and American coasts of this harsh and demanding world.[8]

We know little about climatic fluctuations in the Bering Strait during the Medieval Warm Period. Southerly winds sometimes prevailed for periods of several decades, or even centuries, during periods of warmer climate. But colder winters brought strong northerlies and ferocious storms. So the pattern of settlement shifted, favoring north- or south-facing coasts according to changing climatic conditions. Warmer conditions between 1000 and 1200 did mean that the open-water season was longer, with more narrow channels—"leads"—through the pack in spring where hunters could stalk sea mammals. The climate was still savage, colder in some places, warmer in others, with greater storminess off one coast, calmer conditions off another. A longer ice-free season and more open water allowed people to move more freely, to hunt sea mammals over wider ranges, and to trade with greater ease. There were no universal benefits from warmer conditions as there were in medieval Europe. Each year was different, some with severe ice conditions, others with open water for months on end. To thrive in this challenging environment required a level of adaptability and opportunism almost unmatched in the ancient world.

In many ways, a degree or two of climatic warming was irrelevant to people who dwelled in a world where cold was the norm, where ice and relentless winter were part of daily existence. Unlike the Norse, the native peoples of the Arctic had inherited adaptations to extreme cold from remote Stone Age ancestors, and they passed those skills from one generation to the next. They were adaptable enough to be immune to even quite large climatic fluctuations. Unlike the Inuit, the seagoing Norse knew nothing of Arctic living and were at the mercy of ice conditions in the North Atlantic. Their western voyages might never have reached Greenland and the Canadian Archipelago if it had not been for the centuries of warming.

Bering hunting groups developed a remarkable expertise with skin kayaks and hide boats. Eskimo men, like the Aleuts to the south, started paddling such craft practically from infancy. They protected themselves from the cold by wearing their kayaks like skin garments, but the full effectiveness of such watercraft depended on the development of hunting weapons much more capable than simple thrusting harpoons. Sometime during the first millennium B.C., Bering Strait hunters developed revolutionary sea mammal hunting technology that relied on toggle-head harpoons. Just as iron plows had revolutionized agriculture in Europe, so the toggle-head harpoon changed life in the ancient Arctic. (The heads were originally made of ivory, of course, not iron.) The conventional, line-mounted barbed harpoon head detached from the shaft when it struck its prey. But the head often slipped out of the wound when the quarry dived and struggled. In contrast, a toggled head, which also detached from the shaft when it hit its quarry, had a swiveling tip that toggled under the skin and blubber of its prey so that it could not be detached by violent movement or contact with ice. The new harpoons were especially successful against whales and larger sea mammals when used from light skin kayaks and umiaks, large skin boats.[9]

As the centuries passed, the ivory-tipped harpoons became ever more elaborate, often with fine decoration. The same material provided mouthpieces and plugs for the harpoon-line floats of inflated sealskin that helped the crews to recover larger sea mammals when they died. Ivory was good, tough material for harpoon heads, but no substitute for exotic and highly prized iron, which first reached the Bering Strait region from Asia about two thousand years ago (the precise date is uncertain).

By A.D. 900, whale hunting had developed to a fine art, carried out by teams of expert boatmen and hunters in umiaks. They would pursue

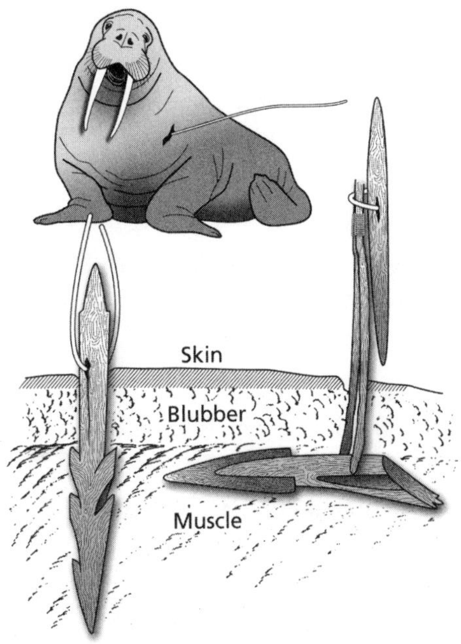

A toggle-headed harpoon penetrates the prey, then detaches from its foreshaft (an intermediate link to the main harpoon shaft) and toggles sideward, causing internal bleeding and making it nearly impossible for the animal to shake the harpoon loose. The head is attached to a line and a float, so the hunter can track a wounded beast.

the migrating whales through narrow ice leads in spring and in open water in the fall. In warmer summers, such as became more common toward the end of the millennium, open-water conditions were such that captains could follow whale migrations over long distances through the strait and along the Arctic Ocean shore to the east. Contact between west and east appears to have become more regular during the warm centuries, just as the Norse arrived off Baffinland.

EKVEN VILLAGE, EAST Cape, Siberia, A.D. 1100. A smoldering fire fills the house with smoke that hovers in the whalebone rafters above the two whaleboat captains. They talk softly, the one a local man, the other from across the strait, from a summer village now named Ipiutak near what is now Point Hope, a place with powerful supernatural associations.[10] The visitor, having taken advantage of the fine late-summer weather to cross the water, arrived a few hours ago. He timed his visit well, for a strong southerly wind now whips the strait into a frenzy. Some days will pass before he can move on to his next destination to the south.

The two men are relatives, both skilled whale hunters, the visitor also a revered shaman known far and wide for his spiritual powers. His men have unloaded a cargo of caribou skins, now piled by the hearth. The host reaches into a sealskin bag and brings out a fine iron spear point, which shines brightly in the firelight. The easterner examines it closely, runs his finger over the sharp point. He shakes his head and the bargaining begins. By the time it ends, a knife, some more fine points, and a lump of dark ore lie on the ground between the captains. Each side has driven a hard bargain, but both are satisfied, for caribou skins, and especially iron, are in short supply.

What remains of Ekven lies on the Bering Sea coast of Chukotka, south of Cape Dezhnev (East Cape) in extreme eastern Siberia, the closest point to Alaska, which is readily visible from the shore on a clear day.[11] A cemetery lies atop a hill a few hundred yards behind the settlement. The remains of more than three hundred people have come from the burial ground. The grave goods tell us that there were sharp social distinctions within Ekven. Three tombs contained two thirds of the harpoons from the cemeteries. One skeleton lay with a remarkable inventory of seal- and bird-hunting equipment, including ten harpoons, spear points, ice picks, and wooden hat accessories. There were multiple burials, too, perhaps of retainers or slaves, some of them perhaps sacrificial victims.

The archaeologist Mikhail Bronshtein has studied the elaborately decorated bone and ivory tools from the cemetery and believes that the fantastic animal-like motifs and other elaborations are signs that distinguish different communities and kin groups from far and wide across the Bering Strait. Some of the graves were those of village residents. Perhaps others were of visitors such as traders, or of wives brought in from other communities. An elaborate web of political and social interconnectedness seems to have linked Bering Sea communities over long distances, all of them living off the maritime resources of the Arctic seas. We can imagine the factionalism and sporadic violence that marked these societies, the quarrels and petty disputes that assumed menacing proportions during the long months of total winter darkness. They competed and fought over trade routes and hunting territories, over perceived insults that festered for generations.[12]

When Bronshtein pored over the Ekven artifacts, he discovered that the lines and tracery on the bone and ivory artifacts were so fine and intricate that they could only have been made by cutting and engraving tools with iron working edges. He believes that most of the ivory, antler, and wooden artifacts of the Old Bering Sea culture were fashioned with iron tools, so this was a vital material for the artisans of two thousand years ago and later.

At first, Bering Strait iron came from trade fairs in the interior, perhaps, as well as down trade routes along the coast. Strategically placed villages like Ekven were in a good position to control all kinds of trade, especially that in valued commodities like iron. We do not know how much iron passed from hand to hand across the strait, but large quantities were probably involved. This could account for the rapidity with which Eskimo groups accepted iron technology and artifacts after later European contact. Iron was a known, highly useful, and valued commodity long before the Russian explorer Vitus Bering sailed into the strait that bears his name in 1728.[13]

By the time conditions warmed slightly in the Arctic, the highly competitive societies of the Bering Strait had developed an insatiable hunger for iron. Almost all of it came from Asian sources. Then, gradually, a handful of iron objects may have come from the opposite direction, from sparsely inhabited lands thousands of miles away in the Canadian Archipelago and Greenland.

EAST OF THE Yukon, a vast tract of low-lying, undulating territory stretches eastward to the distant Atlantic Ocean. A thick ice sheet blanketed this glacier-scoured, rocky land as recently as fifteen thousand years ago. Hudson Bay is the dominant landmark, little more than a shallow basin. The barren terrain of the ice-clogged Canadian Archipelago lies north of the mainland, separated by a short distance from Greenland with its vast ice cap. Throughout this savagely cold world, there are rarely more than three frost-free months a year. Even during warmer centuries, the permafrost would have persisted, covered by bogs and swamps during the summer months, so that travel by land would have been difficult at best. These hardships would have been compounded by swarming mosquitoes. Vegetational cover is sparse, but caribou and musk ox could be hunted, also small fur-bearing animals and seabirds. But the rich resources of the Arctic Ocean provided food aplenty for the tiny primordial Tuniit populations that moved east from Alaska along the shore and into the archipelago after about five thousand years ago.

The Tuniit were tough, resourceful folk who survived in some of the harshest environments on earth with the simplest of technologies.[14] The Canadian archaeologist Moreau Maxwell once described what life must have been like in their tiny musk-ox-hide tents, equipped with hearths, in the depths of winter in about 1700 B.C. The thick scent of seal oil lamps inside the tents would have pervaded everything. "The bitter winter months might have been spent in a semi-somnolent state, the people lying under thick, warm musk ox skins, their bodies close together, and with food and fuel within easy reach."[15] Everyone kept trips outside to a minimum and basically hibernated.

Tuniit culture developed slowly over many centuries. These Dorset people, as archaeologists call them, were constantly on the move, pushing northward in warmer centuries, retreating in the face of colder conditions. The people were, above all, sealers, who also took caribou. During centuries of colder climate, they developed new ice hunting and fishing methods that allowed them to acquire food in midwinter, rather than semihibernate as their predecessors had done. They had but the most basic of spears and no bows and arrows, nor did they have the sophisticated boats or toggle harpoons of Bering Sea people. Their hunting relied on expert stalking and infinite patience that enabled them to approach their prey at close range, then kill it with a spear thrust. Such weapons were valuable for ice hunting through holes in the pack, especially after A.D. 1000 and during the warmer centuries, when the hunters started using weapons tipped with hammered, pure, and extremely rare meteoric iron from the Cape York region of northwestern Greenland, obtained from a meteor shower that fell to earth at least ten thousand years ago.[16] Other groups exploited the native copper outcrops in the Coppermine River region of the central Arctic.

Both meteoric iron and native copper had major advantages over bone and ivory—weapons tipped with metal were tougher and more lethal—and such metal was accordingly precious. Judging from measurements of the slots in abandoned bone tools, precious iron was recycled again and again. At least some of this iron was in the hands of whaleboat skippers. Over 46 percent of the slotted bones from a house with whaling gear at a Thule settlement at Qariaraqyuk on Somerset Island once held metal blades. In a smaller house nearby, 9.6 percent of the tools had held metal blades; all of these artifacts were used for hunting on land. Small quantities of iron slowly passed from hand to hand over enormous distances, in some cases as much as 372 miles (600 kilometers) from their source. In time, some of the ores, or artifacts made from them, may have traveled westward to iron-hungry groups around the Bering Strait.

There must have always been some contacts between the Bering Strait and points east, but it was during the Medieval Warm Period that these connections increased significantly.

LIKE THE NORSE, entrepreneurial canoe skippers in the Bering Strait were both insatiably curious and hungry for new trading opportunities. Warmer conditions after 1000 would have brought more weeks of ice-free conditions, also wider ice leads along which skin boats could pass safely to hunt sea mammals and migrating whales. With more plentiful food supplies, local populations may have increased, which in turn may have caused some whale captains to strike out for new hunting grounds. And with more favorable ice conditions, umiaks could follow bowhead whales in open water and through wide ice-free passages as they migrated eastward along the shores of the Canadian Arctic and into the archipelago. Bowheads, *Balaena mysticetus*, are Arctic right whales with large bow-shaped heads that form up to 40 percent of their body length. They live near the surface and move in small groups during spring and summer and in large groups during the fall.

In 1921–24, a Greenlandic scholar, Knud Rasmussen, led an expedition by dogsled from Greenland to Alaska, studying Inuit groups and excavating archaeological sites along the way. To their surprise, the archaeologists, headed by Therkel Mathiassen, unearthed a quite different culture from that of the living Inuit. They identified the hitherto unknown society of a thousand years ago from abandoned houses near Thule in northwestern Greenland. Soon afterward, they found similar sites over an enormous strip of the Arctic from the Davis Strait to northern Alaska. These Thule people were the sea mammal and whale hunters who had moved across the High Arctic during the Medieval Warm Period, hunters so successful that they dwelled in permanent winter communities of stone and turf houses roofed with whale jawbone beams.[17]

The Thule migration of A.D. 1000 soon passed into the scholarly literature as a movement of whale hunters who had rapidly traveled east along the Arctic coast from the Bering Strait in pursuit of bowheads, which thrived in the more open waters during the warmer centuries of the Medieval Warm Period. In reality, the events behind the migration may have been much more complex, involving not only whaling but a quest for iron.[18]

To what extent warmer conditions played a role in the movement of Thule people eastward from Alaska is unknown. There are some indications that the two warmer centuries also brought strong north winds and many storms. But whatever the conditions, the Thule and their Bering Strait ancestors were more than capable of surviving comfortably, of adjusting effortlessly to greater warmth or cold. Whether it was iron or whales that took small numbers of these people thousands of miles, we cannot be sure. Certainly whales were a critical staple, and remained so. But the real lure may have been meteoric iron from Cape

York, and also outsiders from over the ocean who apparently possessed the precious ore in abundance. In an Arctic world where people covered long distances and where intelligence about ice conditions, whale migrations, and sea mammal rookeries was all-important, we can be sure that the Thule, like the Tuniit, had heard stories of mysterious *qadlunaat*, blue-eyed strangers from over the sea, who used iron weapons and had plenty of metal, sometimes even willingly trading it.

The plot thickens when we realize that the earliest eastern Thule sites lie along the extreme High Arctic and icebound coasts adjacent to Cape York's meteoric iron deposits. These early settlements yield not only meteoric iron but also fragments of iron and other Norse artifacts that can only have come from Greenland. Furthermore, the artifacts from these Thule settlements are identical to tools from communities around the Bering Strait. The Canadian archaeologist Robert McGhee and others believe this may be a sign that people from the Bering Strait region moved rapidly across the north to the Cape York region in an effort to gain control of the sources of iron at a time of warmer climate and perhaps more favorable ice conditions. Radiocarbon dates from the early Thule sites in the east hint that settlement began sometime during the twelfth or thirteenth centuries, at a time when Norse settlements in Greenland were enjoying considerable prosperity, when the summer pack frontier lay far north of Iceland and voyaging conditions in the North Atlantic were relatively easy during the summer.

Once settled in the east, Thule groups spread gradually through the entire eastern Arctic. Their oral traditions recount how they killed or drove away the Tuniit as they gained control of iron sources. By the thirteenth and fourteenth centuries, the remaining Tuniit settlements were abandoned. At some point, too, some Thule bands moved southward from the northwest, came into contact with Norse communities, and coexisted with them. Neither side made an effort to displace the other, for they had commodities to offer one another that could not be obtained any other way.

We can imagine the cautious approach to land as the crew row the *knarr* into the ice-free bay. They keep their bows and swords close to hand as three kayaks approach the slow-moving ship. The paddlers gesture toward shallower water close inshore where the Norse can anchor safely. As soon as the anchor is down, the three Inuit secure alongside and clamber aboard. They are without fear, for they have traded with this ship before. Presents are exchanged—some lengths of brightly colored wool and a fine walrus tusk. The youngest Inuit fingers the iron nails that secure the planks and gazes in amazement at the iron swords in their sheaths. He has never seen so much of the precious metal before.

The bartering moves slowly when the skipper and a few watchful rowmen come ashore in their small boat. The Inuit lay out rows of walrus tusks in front of their winter houses. For their part, the Norse untie bundles of wool and iron bits. Not new weapons or spears, but old boat rivets, fragments of chain mail, handfuls of nails, and fragments of metal barrel straps, all discarded at home or recycled while shipbuilding the previous summer in forested Labrador, but of immense value on this side of the Davis Strait. Once the Norse have departed, the people convert some of the iron into spear and harpoon points, but much of it is simply cherished as a valuable, exotic material.

After several days, the Norse depart with a load of ivory, leaving behind them some iron objects, an old steel helmet, and bolts of fine wool woven during the winter in Greenland. For as long as anyone can remember, the *qadlunaat* have arrived in summer, not every year, but when ice conditions permit, and without warning. Over many generations, the Inuit have come to depend on the trade for supplies of the most precious of all commodities: iron. Presumably, they have stockpiled walrus ivory against these rare visits.

A few Norse artifacts have come from native settlements on Ellesmere Island in the High Arctic.[19] The Ellesmere artifacts include non-native copper and iron, fragments of chain mail and carpenter's tools, also boat nail rivets, fragments of woolen cloth, and a few carvings that give impressions of Norsemen. There are even some reworked bottom sections of wooden casks.

Excavations in Inuit dwellings at Nunguvik on Baffin Island tell us even more. They include strands of yarn identical to some wool fragments found in the Western Settlement, the northernmost Norse community in Greenland. There are fragments of pine wood dating to the late thirteenth or early fourteenth centuries. Pine does not come ashore as driftwood here. Two pieces bear holes with what appear to be rust stains from iron nails. The finds from this site are thought to be evidence of direct contact between Norse and native peoples, rather than objects passed from hand to hand over long distances. Nearly 600 miles (1,000 kilometers) farther south, two sites on southern Baffin Island have yielded Norse cordage and short lengths of yarn.

The dispersion of artifacts found over an enormous area of the sparsely inhabited Canadian Arctic testify to at least sporadic contacts between the Norse and the Inuit. The earliest reference to such contacts comes from a twelfth-century text, the *Historia Norvegiae*: "Beyond Greenland, still farther to the north, hunters have come across people of small stature who are called Skraelings. . . . They do not know the use of iron, but employ walrus tusks as missiles and sharpened stones in place of knives."[20] How far west some of these Norse artifacts eventually traveled, we do not know. But there is absolutely no reason why a handful of them could not have reached the Bering Strait during the warm centuries.

At the time, the Inuit, who were expert walrus hunters, occupied the coasts and islands of Arctic Canada, as far as the northwestern corner of Greenland north of Melville Bay. This region, named Helluland by none other than Leif Eirikson, abounded in walruses, whose ivory was a precious trade commodity for the colonists. Greenland was too cold, even in the warm centuries, for cereal cultivation. So the settlers relied on a dairy economy, growing hay for winter fodder, also on fishing and sea mammal hunting, especially when cooler temperatures descended on the north during the thirteenth century. In 1262, Greenland, like Iceland, became a tributary of Norway, but the real link between the Greenlanders and their home country was the church.[21] The first bishop to live in Greenland arrived in about 1210 and established his residence at Gondar in the south. For generations, the Greenlanders paid tithes to the Norwegian church in valuable commodities—local cloth woven from sheep wool, Arctic furs, live falcons for the royal sport of European and Islamic kings, walrus-hide ropes for ships, and, above all, narwhal and walrus ivory. In 1327 alone, church officials reported a special Crusade tax of about 1,400 pounds (650 kilograms) of ivory, which would have required the killing of some two hundred walruses.

The compelling demands of church tithes took the Norse far north and brought them in contact with the native peoples of Arctic Canada. Over many generations, complex trading relationships developed between the Norse and their indigenous neighbors, fueled by two commodities—walrus ivory and iron. With a tithe that required more than four hundred walrus tusks a year, the Norse needed far more ivory than they could ever obtain around their settlements. Their Inuit contacts were at best sporadic, but they were of mutual benefit. The hunters were coastal people, who tended to stay on the outer shores, far more interested in trade than in displacing the colonists from their farms. It was not until a series of intensely cold winters and unusually cool summers between 1340 and 1360, well documented in Greenland ice cores, that the more northern settlers abandoned their farms and moved southward to join relatives in the more hospitable environment of the Eastern Settlement. With the abandonment of the northern settlements in the face of increasing cold, the ivory trade collapsed. Since ivory was the Greenlanders' main source of wealth, they must have found it increasingly hard to maintain any economic relationship with Europe. The payment of church tithes ceased as ice conditions worsened and ties with Norway ended. By 1370, the practice of sending an annual trading ship from Norway to Greenland had ended. The last bishop died at Gondar in that year and was not replaced.

The Inuit had also become increasingly dependent on trade with their neighbors for iron. As Greenland became more isolated from Europe, demand for ivory virtually ceased, forcing the Inuit to be more aggressive in their dealings with the Norse. They moved southward in the absence of the settlers, looting abandoned farms for metal. Reluctant as the Norse were to adopt native hunting practices or technology, the presence of indigenous hunters deprived the colonists of access to critical seal hunting and fishing grounds at a time when such resources were increasingly important for their survival.[22] By 1450, the Norse colonies in Greenland were deserted and the fleeting contacts between two very different worlds fostered by warmer temperatures ceased. Only the Norse epics and treasured oral traditions preserved memories of an era when native Americans and Europeans met for the first time.

The Norse and Inuit, like the inhabitants of northern Europe, found their lives made easier (if never easy) by the warming of the climate. Food was more abundant, and new technologies made the age-old labors of the farmer and the hunter more productive. From the Arctic to North Africa, with greater ease of transport, cross-cultural contacts allowed some of these technologies to be shared across great distances. Elsewhere on the planet, however, the rise in temperatures was not so benign. The great warming brought bounty to some areas, but to others, prolonged droughts that shook established societies to their foundations.

CHAPTER 6

The Megadrought Epoch

In the beginning there was no sun, no moon, no stars. All was dark, and everywhere there was only water.

—Maidu creation legend, California[1]

YOU SWEAT EVEN WHEN SITTING under the deep shade of the rock shelter. A vast panorama of desert landscape lies before you—arid, heat-blasted mountain peaks, a pale, dusty blue sky overhead. The heat shimmers above the desert floor, over dunes and dry streambeds, the sparse shrubby vegetation grows near the ground. The sun is moving to the west, but the air is still, the silence complete. No wind sloughs through the scrub or causes sandy williwaws to traverse the searing plains. Day after day, one rises with the dawn and takes refuge from unrelenting sunlight well before midday. And it is only early June, with weeks of even greater heat still ahead.

As soon often happens, your mind goes back deep into the past, in this case to the generations of foragers who once visited this place and looked out over the same arid vista. Only a handful of visitors would arrive each time, perhaps a dozen men, women, and children, the adults thin, agile, and wrinkled, as if parched by the desert sun. The women would light a fire as the sun approached the western horizon, while the men would scout for jackrabbits feeding on the banks of a sluggish-moving nearby wash. Back in the shelter, the women would grind some piñons from a precious store carried in a deer skin. The soft scraping of the milling stones was a familiar evening sound, part of the unending quest for food that kept the band on the move nearly all the year. The meal is sparing at best. No one is hungry, but edible plants are scarce. Even rabbits are hard to find after a very arid year.

The American West is landscape on a grand scale, the stuff of legends, of John Wayne and classic Western movies. From 40,000 feet (12,000 meters), you gaze down at the dry terrain seemingly hour after hour, at a semiarid world that's larger than life. Tough wilderness country breeds legends and stereotypes of hard-bitten men and resourceful women, the characters beloved by Hollywood. Reality was, of course, much more complex, but the sheer scale of the western landscape dwarfs humanity and leaves one in awe of the hunters and plant gatherers who thrived in this inhospitable world for thousands of years before the first cowboy tended cattle here. Europe may have enjoyed bountiful harvests and the Norse voyaged more freely in the North Atlantic, but, like Eurasia and West Africa's Sahel, the American West suffered under megadroughts.

THE GRAY LIGHT of a clear sky before dawn spreads across a dry lake bed. The men crouch low among the shrubs on the dry floor of a huge, rapidly shrinking lake in what is now California. This is the driest year they can remember. The lake has shriveled before their eyes through months of great heat, leaving extensive sand flats in its place. They and their neighbors have camped where water once stood. The men have moved into position well before sunrise, using boulders and the streambed to stay out of sight. Each hunter carries a bow and quiver of arrows, eyes casting left and right for a sight of the deer feeding in the cool of morning. Most likely, the beasts will be close to a small water hole at the lake's edge. Two of the young men exchange glances as they spot a feeding buck. They move softly, stalking their quarry, alert for the slightest whisper of a morning breeze that could carry their scent. With infinite care, they approach ever nearer to the feeding deer. After half an hour they are within range. Suddenly, their prey looks up, sniffing the air. Perhaps he has caught a whiff of human scent. The men freeze, weapons still. Minutes pass as the buck scans his surroundings. Finally, reassured, he resumes feeding. The men raise their bows, slowly notch their stone-tipped arrows in place. They crouch for the best shot, but someone's foot taps a stone on the ground. The startled deer is instantly on the move. Two

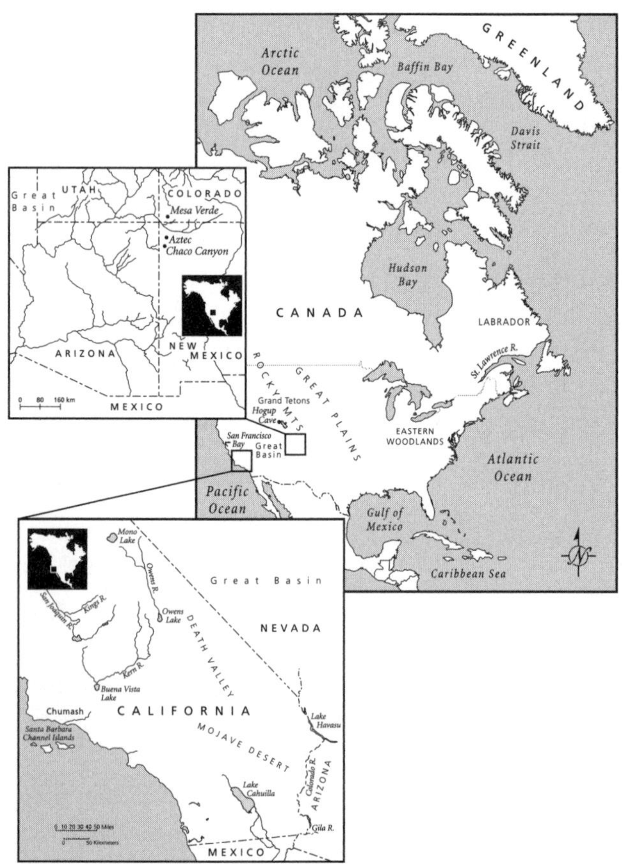

Locations mentioned in chapters 6 and 7. Some minor places are omitted for clarity.

arrows fly, but miss, skittering harmlessly across the lake bed. By now the sun is rising fast, so the hunt will have to wait for evening or another day.

Owens Lake on the eastern flanks of the Sierra Nevada Mountains in eastern California provides telling evidence of the epochal droughts that descended on the West between A.D. 900 and 1250. The lake once covered over 115 square miles (300 square kilometers) at the mouth of the Owens River and held water continuously for at least 800,000 years. (Owens Lake was more than 250 feet [75 meters] deep until the Los Angeles Department of Water and Power diverted the streams that fed it in 1913 and it became a large salt flat.) The mountain runoff that flowed down the river varied dramatically from year to year in centuries-long, decadal, and even annual cycles of unusually wet and dry years. In drier periods, cottonwoods and Jeffrey pines grew in the moist soils of the receding lake bed. When wetter years brought rising water levels, the trees drowned. For years, the dead trunks and branches would stand above the water, but they would eventually disintegrate, leaving only the stumps rooted into the now covered lake bed.

The geographer Scott Stine has spent much of his career studying these once flooded tree stumps, which were exposed by receding lake waters in drought years. By radiocarbon-dating the outermost tree layers and then counting the tree rings from the stumps, he has reconstructed a precise chronology of droughts and wetter periods during the Medieval Warm Period that are startlingly consistent over a wide area of the American West.[2]

Stine's research began during a major drought during the 1980s, when dry conditions and heavy water demand in Los Angeles caused Mono Lake, the northernmost catchment of the Los Angeles aqueduct, to drop by more than 49 feet (15 meters). He collected samples from numerous stumps, radiocarbon-dated them, and found that there were two generations of trees and shrubs that grew in the lake during the Medieval Warm Period. The first generation perished when the lake rose some 62 feet (19 meters) around A.D. 1100. The rise came during a brief very wet cycle when the rainfall was higher than in any year in modern times, and the fourth highest of the past four thousand years. But plentiful rainfall gave way in about 1250 to an intense drought phase, which lasted for over a century. The lake fell precipitously and a second generation of trees rose in the newly exposed bed.

The Mono Lake tree stumps chronicled a Medieval Warm Period marked by extreme rainfall swings within a century or less. Intrigued, Stine now turned his attention to Lake Walker to the northeast, a body of water nourished by two Sierra rivers. The western of the two flows through a narrow canyon studded with large submerged pine stumps. Since the canyon is very narrow and lateral movement of the river is restricted, it seems certain that these trees flourished at a time of greatly reduced flow, for pine roots cannot tolerate more than brief periods of inundation. The tree stumps documented a very low level around 1025, when the water stood more than 131 feet (40 meters) below today's shoreline; this was followed by a brief wet cycle, then another drought, the chronology the same as that from Mono.

Owens Lake has also provided evidence of severe medieval droughts at the same time. Hunter-gatherers wandering the dry lake bed between A.D. 650 and 1350 left distinctive stone projectile points behind them at a time when the lake was severely desiccated. Radiocarbon dates from a nearby, and contemporary, root of a once flourishing shrub narrowed the occupation down to the time of the first major drought recorded at Lake Walker. Mono, Walker, and Owens lakes all record the same medieval dry cycles. The first began before A.D. 910 and lasted until about 1100. The second commenced prior to 1210 and ended in about 1350. How severe was the drought? Stine used a modern baseline, that of the six-year California drought that began in 1987, when Sierra Nevada runoff was only 65 percent of normal. Despite prolonged dry conditions, the lakes never fell as low then as they had in earlier times. To account for the desiccation experienced by Owens Lake, for example, inflow to the lake must have dropped to between 45 percent and 50 percent of modern amounts.

Stine's droughts were so severe that one can trace them over large areas of the West. They turn up in tree rings from the White Mountains of eastern California, where long-lived bristlecone pines are sensitive to temperature or rainfall changes. One record produced evidence that the period 1089 to 1129 was the wettest cycle of the past thousand years. In the southern Sierra, a thousand-year record from foxtail pines and junipers revealed the same two savage droughts, with the four warmest periods of the past millennium occurring between the tenth and fourteenth centuries. The warmest stretch of all was between A.D. 1118 and 1167.[3]

Evidence of the droughts extends as far north as east-central Oregon, into the Rockies, and into the adjacent Great Plains. Upright tree trunks stand in 78 to 98 feet (24 to 30 meters) of water in Jenny Lake in Grand Teton National Park. The divers who inspected them even found a raptor nest in the branches of one of the submerged trees. Outer wood from one stump has been radiocarbon-dated to about A.D. 1350, virtually contemporary with the dead stumps in Sierra lakes, as if water levels were also reduced in this vicinity. The size of pocket gophers in northern Yellowstone National Park was the smallest in three thousand years as small rodent populations fell rapidly in the face of aridity.

These major droughts occurred because the winter jet stream over the northeastern Pacific, with its associated storm tracks, stayed well north of California and the Great Basin. A classic instance of the same phenomenon came during the 1976–77 winter rainy season, which was the driest over much of California since records began. Alaska, on the other hand, enjoyed the wettest ever known. The same jet stream pattern persisted over the American West during much of medieval times, while Alaska was exceptionally wet. We know this because geologists have recovered newly exposed tree stumps under retreating glaciers in Prince William Sound and dated them to between A.D. 900 and 1300—the Medieval Warm Period. The ice sheets had advanced during the cycle of wet years, as they do today during wet winters.

Stine's observations, and those of others, have received strong validation from a grid of no less than 602 tree-ring sequences dating back as far as two thousand years.[4] For the first time, climatologists have been able to compile a grid of drought data from the entire region, which uses two indexes, the Palmer Drought Severity Index, a well-established way of measuring fluctuations in wetness and aridity, and a Drought Area Index, which counts the number of locations on a grid that exceed an arbitrarily established threshold of the Drought Severity Index. These calculations enabled the researchers to put current twenty-first-century droughts in the West into longer-term perspective. The four driest periods centered on A.D. 935, 1034, 1150, and 1253, all of them within a four-hundred-year interval of overall aridity, which coincides with the Medieval Warm Period. After 1300, there was an abrupt change to persistently wetter conditions, which lasted for six hundred years, then gave way to today's drought conditions in the West. None of today's droughts, which last as long as four years, approach the intensity and duration of the Medieval Warm Period droughts. The latter were so severe that there is talk of a "megadrought epoch" a thousand years ago.

It's the prolonged nature of these dry periods that distinguishes the aridity of the Medieval Warm Period from today. How did these droughts occur, and, above all, how did they persist for so long? The researchers believe that unusually warm conditions contributed to the occurrence of more frequent, persistent droughts, which were caused in part by great evaporation and reduced moisture levels in the soil. Climatic fluctuations in the Pacific have a considerable effect on rainfall in the West, especially the effects of El Niños and their opposites, cool and dry La Niñas (the El Niño–Southern Oscillation [ENSO]), and a phenomenon known as the Pacific Decadal Oscillation (see sidebar), which can cause drought in the West.[5] Furthermore, increased northern hemisphere temperatures during the twentieth century and unusual warming of the western Pacific and Indian oceans have contributed to drought formation over middle latitudes in the northern hemisphere. The same effects presumably occurred during the Medieval Warm Period. In other words, large-scale warming at a more global level contributed to the megadrought epoch.

There are other factors, too. Increased upwelling of cold water in the eastern Pacific is apparently linked to heating of the tropical atmosphere. The enhanced upwelling in turn promotes the development of La Niña–like cool, dry sea surface temperatures over the eastern Pacific—and drought in western North America. As we shall see in chapter 10, such conditions were indeed persistent during the warm centuries. During the period 1150 to 1200, volcanic activity was reduced globally and sunspot activity was high, both of which contributed to a cooler, La

The Pacific Decadal Oscillation

The Pacific Decadal Oscillation (PDO) is a long-term fluctuation of the Pacific Ocean. During cool phases, an area of lower than normal sea surface heights and temperatures is present in the eastern equatorial Pacific. (As the ocean warms, it expands and the surface becomes higher.) An equivalent wedge of warmer, higher than normal sea surface heights and temperatures connects the north, west, and southern Pacific. During warm phases, the eastern Pacific warms up and the western parts of the ocean cool. These changes in cold and warm water alter the path of the jet stream, which flows farther north during cool phases, thereby reducing rainfall in the west. PDO phases wax and wane about every twenty to thirty years. El Niños and La Niñas superimpose themselves atop these long-term fluctuations. It appears that we are currently entering a cool phase, which means less rainfall over much of western North America for two or three decades.*

* See http://sealevel.jpl.nasa.gov/science/pdo.html.

Niña–like state across the eastern tropical Pacific, a condition that brought drought to wide areas of the Americas.

THE MEDIEVAL DROUGHTS of the Sierra were among the most severe of the past four thousand to seven thousand years, far more savage than those we complain about today. What, then, were the impacts on the human populations of the day, when at most a few hundred thousand people lived in the Great Basin and along the Oregon and California coasts? The harshest conditions occurred in the arid interior, especially in the Great Basin and Mojave Desert, where human populations were always tiny, even near the swamps and lakes that supported larger numbers.

The Great Basin covers over 360,000 square miles (1 million square kilometers) of the desert West between the Rocky Mountains and the

Sierra Nevada—parts of California, Oregon, Utah, and Idaho, and nearly all of Nevada. This is a world of great environmental diversity, of high mountains and intervening valleys, where the topography varies dramatically and environmental zones are stacked vertically. The arid southwestern portions of the basin support the sparsest vegetation, while the mountains are somewhat wetter and have more complex climatic regimens. Fortunately for hunters and gatherers, there are some significant lakes and wetlands in valley bottoms, which were once magnets for the small numbers of people who lived in the basin. Most of this vast area receives almost no rainfall, and the amounts vary greatly from one year to the next. In ancient times, this meant that the plant food supply could be as much as six times more plentiful in a wet year than in a dry one, although the word "wet" is really a misnomer in such an arid world. To survive here required unlimited patience, opportunism, and a very broad diet. To eat one food at the expense of all others was to invite disaster. The secrets of survival were constant mobility and an intimate knowledge of dozens of edible plants.[6]

Climatic conditions were never particularly stable in the Great Basin, for major changes in local environments could occur even within the brief compass of a single year. Food supplies were patchy, often confined to relatively productive locations, such as small wetlands, separated from other such places by much more barren landscapes. While a few bands dwelled in exceptionally rich lakeside or marshy environments, people in drier areas lived in tiny family groups. They were constantly on the move, subsisting off seasonal foods at widely separated locations. Their social life revolved around intelligence gathering, around "mapping" the changing availability of different foods throughout the year. Every western hunting band spent a great deal of time acquiring information, from fellow kin in other bands, from visitors, from people bartering glasslike tool-making stone from far away.

People living in such arid environments had few options, with the result that their basic way of life survived virtually unchanged over thousands of years. At Hogup Cave in central Utah, over 13 feet (4 meters) of sporadic occupation chronicle more than eight thousand years of occasional visits and a conservative lifeway that changed little over the centuries.[7] Whenever they stayed at the cave, local bands relied heavily on pickleweed, a low-growing succulent that thrives on the edges of salt pans and dried-up lake beds. They gathered all kinds of edible plants in fiber baskets (these are preserved in the cave deposits), then dried or parched the seeds in tightly woven basketry trays, tossing them with some hot embers. Then they hulled and ground the seeds with hand stones and milling slabs. Hogup residents also hunted thirty-two species of small animals and thirty-four of birds with snares, nets, and spears. By consuming a wide variety of foods, they minimized risk of starvation, the cave being one stop on a seasonal round that covered a large territory.

Another seasonal location remained in use for thousands of years, a rock shelter near the former Lake Bonneville (known to archaeologists as Bonneville Estates) in the high desert of eastern Nevada, a virtual inland sea during the Ice Age 18,000 years ago but now long dry. Local bands visited the rock shelter at intervals for over 12,500 years. By 6000 B.C., the visitors ate a wide range of seeds such as buckwheat and wild rye, also piñon nuts. This basic diet continued in use for thousands of years.[8]

Survival in such a landscape depended on diversification, mobility, and intelligence about food and water, a strategy that worked successfully through drier and wetter cycles and through the Medieval Warm Period. Great Basin peoples like the Paiute and Shoshone depended heavily on piñon nuts, a major winter food harvested in late summer and early fall. Nineteenth-century accounts tell us that the people used long sticks to gather green pinecones from the trees; they roasted the cones to release the seeds before parching and grinding. Piñons stored in grass- or skin-lined bags would keep as long as four or five years. Since abundant nut harvests occurred in three- to seven-year cycles, the caches helped tide people over from one season to the next.[9]

Piñons were long a staple in Great Basin life, but the trees are vulnerable to droughts, which bring outbreaks of bark beetles. The insects bore holes in the trees to lay their eggs. Soon the larvae emerge and kill the pines. The U.S. Forest Service estimates that the droughts of 2001 to 2005 have killed an estimated eighty million piñon trees in Arizona and New

Mexico alone, leaving huge swaths of browned forest across the landscape. We have no means of knowing what damage the much severer droughts of a thousand years ago did to piñon forests, but nut harvests must have plummeted for generations during medieval times. The only way one could survive drought under this circumstances was by consuming a wide variety of plant foods, which served as a cushion if the staple failed.[10]

Even in wetter times, population densities were never high. Just like the people of Africa's Sahel, family bands fanned out over the desert in wetter years when there were plants to be found and standing water to drink. In times of drought, as the desert pump pushed them outward, they would retreat to the few places where there were reliable water supplies. Diets varied greatly from place to place. Some groups living by wetlands, like certain groups in Nevada, may have obtained as much as 50 percent of their diet from fish. In contrast, the people living around Owens Lake in California ate almost entirely plant foods. Mobility and a varied diet worked even through the worst of times. The highly mobile Shoshone people of the California desert are thought to have lived in their homeland for between five thousand and six thousand years, a remarkable example of just how flexible hunter-gatherer life can be.

CONDITIONS WERE EVEN severer in the Mojave Desert of the south, which is famous for its high summer temperatures, and was an even more demanding and drier environment for ancient humans than the rest of the Great Basin. During the 1890s, the anthropologist David Prescott Barrows studied the foraging practices of the Cahuilla Indians, who flourished in the Mojave for thousands of years. He found the same strategies for coping with dry conditions as elsewhere in the desert West.[12] Even in wetter times, everyone lived in places with the coolest temperatures and where water was most plentiful. The Cahuilla diet was broadly based. Each band exploited several hundred plants for food, manufacture, or medicine; the Cahuilla harvested six varieties of acorns in fall and relied on mesquite trees, 1.2 acres (0.5 hectares) of which could yield as much as a hundred bushels of beans. Edible cacti, piñon nuts from trees on higher ground, and fan palm dates—the list of plants goes on and on. Eighty percent of all their foodstuffs came from within 5 miles (8 kilometers) of their settlements. But, for all the variety of plants, the unpredictability of the environment left the people in a continual state of uncertainty, as it had their ancestors all over the desert West for thousands of years. Fortunately, the large cushion of edible plants and constant mobility made even severe droughts survivable.

The same strategies of broad-based diet and mobility worked well over thousands of years of climate change, but just how severe were conditions in the Mojave a thousand years ago? Here, too, climatologists rely on an esoteric data source, the tiny middens accumulated by pack rats (*Neotoma*). Between A.D. 600 and at least 1200, the middens contain vegetation from arid surroundings, with few signs of water-loving plants. Furthermore, between 900 and 1300 there are virtually no records of increased spring activity or high lake levels, both characteristic of the Little Ice Age that followed. There was certainly less winter rainfall than in earlier and later times.

Extended droughts in the Mojave after A.D. 800 would have abridged water sources of all kinds and reduced spring discharge. The shallow desert playa lakes, normally magnets for game and waterfowl, would have dried up. Water supplies would have been scarce and widely dispersed, and the risks of traveling to locate them in an arid landscape correspondingly larger. The few archaeological sites that date from A.D. 800 to 1300 lie near major springs and in perennial oases along the Mojave River, where the groundwater was close to the surface even during sustained droughts. Elsewhere, the desert was effectively uninhabitable, except, perhaps, for a few weeks after rare major rainstorms. Life during the Medieval Warm Period was difficult, existence often marginal, but for those adapted to the environment, the droughts were survivable.

Nothing demonstrates the volatility of life in the desert better than the life and death of Lake Cahuilla.[11] Every spring, the Colorado River to the east of the Mojave rose in flood, alternating among channels in its enormous delta. Sometime around A.D. 700, a prolonged shift of the river caused water to flow into the low-lying Salton basin, which filled like a bathtub to a height of about 42 feet (13 meters). At its height, this inland sea was 115 miles (185 kilometers) long, up to 35 miles (56 kilometers) across, and 314 feet (96 meters) deep, one of the largest lakes in North America. The great lake survived for more than six centuries as an overflow for the Colorado, until rising silt levels blocked the entry channel. The lake became a closed basin and dried up within a half century or so.

Thanks to the Colorado floods, Lake Cahuilla remained relatively stable for hundreds of years, its water level varying by 3 feet (1 meter) or so. Geological surveys have shown that the lake was full during most of the warm centuries. The sudden appearance of a huge lake in the midst of an extremely arid landscape was a godsend to local hunter-gatherer groups during a period of persistent drought, but, unfortunately, the water was too saline to drink. Numerous short-term camps flourished along the lakeshore, occupied when people caught fish or trapped and hunted the waterfowl that abounded there.

Throughout the Mojave and the Great Basin, the warmer centuries rendered much of the landscape uninhabitable. Food was less readily available; bands foraged over much smaller areas, only within walking distance of water. Competition for food intensified, and social relationships came under severe stress. Small bands would have coped with reduced nut harvests, families elbowing out their neighbors at the few stands of wild grasses in spring. Words would be exchanged, curses would lead to blows and perhaps even to sneak attacks with clubs and spears. In the end, the group would split up as one party to the dispute would move away to camp with relatives or to found a new settlement. The social environment would have been volatile, charged with anxiety and stoic acceptance of hunger and thirst.

Survival in the arid American West depended on cooperation and, above all, on intelligence about water supplies and food resources scattered over enormous, harsh landscapes. These landscapes were inconspicuously edible living environments when there was water, but utterly devoid of food when drought came and the desert pump expelled animals, people, and plants to the margins. What was for Europe and the

Arctic a time of relative climatic benevolence was for vast areas of the arid American West a time of deprivation and suffering, even among peoples who knew that flexibility and mobility were the elixirs of life. But their environmental knowledge and opportunism helped them survive, perpetuating a lifeway that had survived extremes of flood and drought for many thousands of years in some of the toughest environments on earth.

CHAPTER 7

Acorns and Pueblos

If the drought withered the corn shoots, if the buffalo unaccountably shifted, or if the salmon failed to run, the very existence of people in other regions was shaken to its foundations. But the manifold distribution of available foods in California, and the working out of corresponding means of reclaiming them, prevented a failure of the acorn crop from producing similar effects. It might produce short rations and racking hunger, but scarcely starvation.

—Alfred Kroeber, *Handbook of the Indians of California* (1925)[1]

THE CHANNEL IS GLASSY SMOOTH, the ridges of the snowcapped mountains razor clear against the sharp blue sky of the winter day. The crews paddle in the lightest of southeasterly airs, relishing the calm after the strong northwesterlies of the day before that had followed the rain and snow of a strong winter storm. The ocean slowly turns a gorgeous, roseate pink with the westering sun, a color that lingers long after sunset. Basking sharks move lazily close to the surface, their fins and tails moving imperceptibly in the calm. The weathered, planked canoes glide swiftly over the mirrorlike water, taking advantage of the calm instead of "the very high waves of the Pacific Sea, which, at times, seem to lift them as far as the clouds and at other times to bury them in the bottom of the sea."[2] The paddlers crouch on sea grass pads, paddling steadily, hour after hour, chanting the same canoe song over and over again, as a boy huddled in the bottom of the boat bails with an abalone shell. The echoes of the canoe chant float on the still air as the afterglow of sunset fades and the men come close to land.

The great droughts of the warm centuries in the American West tested ancient strategies of movement to the limit. People died; there was certainly hunger; in some areas, such as southern California, there were startling changes in the nature of society itself, which became more hierarchical, even more despotic. But, through it all, the inhabitants of the West quietly survived. Their traditional lifeways would have continued indefinitely, had not European explorers, then missionaries and settlers, introduced new, totally unfamiliar factors into cultural equations that had thrived for centuries, even millennia.

By no means all western societies lived in desert landscapes. As one moved west toward the Pacific, the varied terrain of what is now California supported some of the most diverse environments on earth—small mountain valleys, local wetlands, and lush valleys with shallow lakes. In some areas—San Francisco Bay and the Santa Barbara Channel for instance—rich maritime fisheries, shellfish beds, and sea mammal rookeries provided so much food that many groups could remain in one place for most of the year if not all. Away from the ocean or permanent rivers and lakes, people relied heavily on wild plants, on widely scattered foods that came into season at different times of the year. Every forager, young or old, male or female, had an encyclopedic knowledge of edible environments, but people were constrained to move constantly to exploit them. This mobility reduced some of their vulnerability to drought cycles, but, with the notable exception of the drought-susceptible piñon, few plant foods could be stored for any length of time. Bands living in some parts of the Great Basin and over much of California came to rely on a surprisingly nutritious staple: the carbohydrate-rich acorn.

Acorns came into use wherever oaks grew in California after about

2000 B.C., at first as a supplemental food. This was a time when the climate was cooler and wetter than today and California's population was rising steadily. Both grass seeds and other long-utilized plants became scarcer as a result. We know there was hunger, because human skeletons dating to earlier than 1000 B.C., before acorns came into widespread use, display many more signs of malnutrition, such as Harris lines (lines that appear in limb bones as a result of temporary retardation of growth) and dental hyperplasia (layers of defective tooth enamel in children) than later burials. The changeover to intensive acorn consumption may well have resulted from overcrowding, more frequent food shortages, and malnutrition episodes. Interestingly, once acorns became a staple, the incidence of dental caries increased dramatically, owing to a carbohydrate-rich diet.[3]

Once the changeover took place, each band readily accepted the extra labor involved in processing and storing acorns, which provided much more abundant and nutritious food. The staple could be stored in quantities so large that many harvesters collected a full year's supply in a few weeks. The only other solution to recurring hunger would have been to start growing maize and beans, just as people were doing in the Southwest. The acorn foragers were well aware of agriculture, but why take the trouble of clearing fields and growing crops when acorns are ready to hand? By the time of Christ, the acorn was a staple of California life from Oregon to the southern deserts.

The acorn did not revolutionize life in the California of a thousand to fifteen hundred years ago. This humble nut merely made food supplies more predictable for people living in a more crowded and territorially circumscribed world. But, during the prolonged droughts of the Medieval Warm Period, heavy dependence on acorns increased the risk of famine, for the acorn harvest was seriously affected by dry conditions.

A.D. 1100. A grove of California oaks on a hot fall day. The bright sunlight casts fretted shadows on the ground during the hectic weeks of the ancient acorn harvest. The band has carried stacks of empty baskets to the trees early in the day. While the older men hunt for deer that feed on the rich mast from the oak groves, young boys leap into the branches and shake the boughs gently. Ripe acorns shower the women and elders. They laugh as they gather the plump nuts into baskets, combing the ground carefully, quickly discarding cracked or rotten nuts, keeping only those that will store well. As the youths climb more trees, the women carry the full baskets back to camp, where they dump the acorns on dry hides. The harvest continues from dawn until dark, for there are only a few days in the year when the acorns are easily harvested.

Early Californians were by no means the first people in history to rely on acorns. Entire villages in Syria subsisted off acorns fourteen thousand years ago. Medieval European farmers consumed bushels of acorns; so did native Americans in the Midwest.[4] As late as the nineteenth century, acorns provided about 20 percent of the rural diet in Italy and Spain. But acorns were nowhere so important as they were in California, with its highly varied environment and endemic droughts. They have excellent nutritional value and are also highly tolerant of storage. Under favorable circumstances, they will keep for up to two years in large baskets or specially built granaries, a priceless quality for people who lived in unpredictable environments. By A.D. 500, thousands of California Indians depended heavily on the immense acorn harvest. At Spanish contact in the sixteenth century, native Californians harvested more than 66,000 tons (60,000 metric tons) of acorns a year—more than the industrialized sweet corn crop in the state today. For many groups, acorns constituted over half the diet. The ancient precaution of a broadly based diet had yielded in the face of abundance.

We know a great deal about acorn consumption, thanks to finds of acorn fragments in numerous archaeological sites, also the grinders and pounders used to process them. Fortunately, too, we have an archive of information about the harvesting, processing, and storage of acorns from both observations of traditional societies and modern-day consumption practices. Balanophagy, acorn eating, flourishes, for many people swear by tasty acorn bread. As a result, we have a great deal of information on this all-important ancient (and modern) food.

Fifteen species of oaks grow from Oregon to California. In good years, California oaks could yield as many as 1,728 pounds/2.5 acres (784 kilograms/hectare), a yield rivaling that of medieval agriculture in Europe and sufficient to support fifty to sixty times more people than lived in California when the Spanish arrived. Unfortunately, however, the harvest fluctuated from grove to grove, even from tree to tree, and most oak species produced a good crop only every two or three years. The ancient foragers, well aware of this, responded by harvesting acorns from different areas. The long storage life of acorns helped compensate for irregular harvests, but acorns have other serious disadvantages. They are a labor-intensive crop, not so much to harvest as to process. Shelling and pounding them is just the beginning, for because of the bitter-tasting tannic acid they contain, they are inedible unless soaked in water.[5] Processing acorns took far longer than grinding seeds, by one estimate about four hours to produce 2.4 pounds (about 1 kilogram) of meal from 5.9 pounds (2.7 kilograms) of pounded acorns. But the effort was well worth it, for the nuts or processed flour could be carried on long journeys or in canoes, as well as used to make nourishing breads and gruels.

By the time the Medieval Warm Period began, around A.D. 900, most forager groups spent more and more of their time in permanent base camps within narrow compasses of territory, in many cases traveling no more than some 5 miles (8 kilometers) from their homes during their entire lives. The small sizes of tribal territories also made each group dependent on others. When droughts reduced the productivity of the environment, kin ties with other groups became very important as a way of acquiring food. Each group had control over local food supplies, such as, for example, acorn groves, fishing pools, or reliable stands of edible seeds. These they exploited more intensively than they would for their own needs, passing food to widely separated communities as part of a complex web of interconnectedness that linked dozens of small villages and the individuals who presided over them.

One could not survive in California's highly localized and varied environments without the help of others. The range of artifacts and commodities exchanged between communities takes one's breath away—all kinds of foodstuffs from acorns and game meat to mollusk flesh, as well as other essentials such as salt and medicinal plants. Asphaltum, for waterproofing baskets, changed hands, as did basketmaking materials and bow wood. Obsidian, a volcanic glass, was highly prized and found at only a few locations, for example Medicine Lake Highlands in northern California. Fortunately, obsidian from different locations has distinctive trace elements that can be identified by spectrographic analysis, so we know that obsidian from Medicine Lake traveled over 50 miles (80 kilometers) from its source.[6] Millions of abalone and *Olivella* seashell beads circulated through California during the warm centuries, as populations rose and territories became smaller. The ancient Mojave trail carried Pacific shell beads as far as agricultural communities in the Southwest. Other bead strings passed from hand to hand deep into the Great Basin. But most trade was purely local, between people who depended on one another for essential commodities in increasingly circumscribed environments.

The prolonged droughts of the Medieval Warm Period played havoc with long-established trading arrangements. Live oaks are susceptible to drought, which reduced acorn harvests considerably over and above the normal interannual fluctuations. We know little about the effects of long-term drought on these trees, which are well adapted to long dry seasons, but seasonal aridity and multiyear dry cycles are different matters. For instance, between 1986 and 1992, a six-year drought killed thousands of oak trees in coastal central California, especially those growing on south-facing slopes, which tend to be drier and with less fertile soils. About 10 percent of all oaks, regardless of species, had perished by 1992.[7] Had not 1993 and later years been wetter, the mortality would have increased dramatically, well above the normal casualty rates from fires, insects, and other causes. Project the 10 percent figure back a thousand years and factor in prolonged, repeated droughts, and the effects on acorn harvests must have been brutal.

Unfortunately, archaeology is an anonymous record of history, so the voices of those who lived through the dry centuries are silent. Only a few shreds of evidence come down to us, including some skeletons from two Yokut Indian cemeteries from the San Joaquin Valley discovered during the construction of the Interstate 5 freeway. The burial places bracket the medieval droughts and were used by people who relied heavily on acorns. Using computer tomography, the biological anthropologist Elizabeth Weiss measured the cortical thickness of femurs, a cumulative dimension that reflects changes in diet and nutrition over a person's lifetime.[8] She found that the individuals buried in the earlier cemetery, which dated to about two thousand years ago had thicker cortical bone and fewer pathologies or traumatic injuries than those from the burial ground dating to between eleven hundred and twelve hundred years ago, the time of prolonged drought. These people had thinner cortical bone, lived shorter lives, and manifested many more injuries, as well as showing evidence of malnutrition.

Acorns were certainly inadequate protection against prolonged drought for people who relied on little else for their diet, but we need another generation of research to fill in the details.

THE SITUATION WAS somewhat different in areas around San Francisco Bay and in the Santa Barbara Channel region, where rich marine environments allowed people to settle in the same place for long periods of time. In the case of the Santa Barbara Channel, natural upwelling close inshore provided rich anchovy harvests, except in occasional El Niño years, when rising sea surface temperatures affected marine productivity. But even in these favored areas, life was uncertain, especially when drought affected drinking water supplies.

The Santa Barbara Channel, south of Point Conception, is another climatic Rosetta stone, site of an exquisitely precise 656-foot (200-meter) deep-sea core that extends 160,000 years into the past.[9] The uppermost 56 feet (17 meters) represent the past eleven thousand years, and rapid sedimentation rates on the sea floor provide fine-grained climatic information. We actually have a record of temperature change at 82-foot (25-meter) intervals over the past three millennia. The core abounds in two species of planktonic foraminifera, one that dwelled close to the surface, the other inhabiting a depth of about 197 feet (60 meters). By analyzing the oxygen isotope content of these organisms and radiocarbon-dating twenty samples of the tiny shells, the oceanographer James Kennett has been able to reconstruct cycles of changing sea surface temperatures within a regimen that was generally reasonably warm, around 54.5 degrees F (12.5 degrees C) on average.[9] From the Kennett temperature curve, we learn that there were warmer sea temperatures between about 2,900 and 1,500 years ago, with a colder cycle following between 1,500 and 500 years before the present, after which the temperature warms until modern times. The coldest sea surface temperatures since the Ice Age fell between 1,500 and 500 years ago, centuries that included the Medieval Warm Period.

Marine productivity in the Santa Barbara Channel rose and fell with cold and warm intervals. During cool periods, the foraminifera show that there was only a slight temperature gradient between 197 feet (60 meters) and the surface, as if the natural upwelling in the channel was very strong and there was constant vertical mixing of seawater. This process was especially intense between 1,000 and 500 years ago. Upwelling, and marine productivity, were dramatically reduced during the warmer cycle between 2,900 and 1,500 years ago. Furthermore, ocean temperatures displayed marked instability between 1,500 and 800 years ago.

The climatic shifts in the Santa Barbara Channel are matched by those recorded on land. Especially low sea surface temperatures appear to coincide with the prolonged drought cycles over much of western North America, especially with those recorded from bristlecone pine tree rings in the White Mountains of eastern California.

The Santa Barbara Channel was home to many Chumash Indian groups, whose ancestry goes back deep into the remote past. When the first European explorers anchored off this coast in the sixteenth century, they reported large villages and dense populations both on the Channel Islands close offshore and on the mainland.[10] For many centuries, we know, the coastal population remained relatively sparse. Then, after about three thousand years ago, trading activity between neighboring communities and between the coast, the offshore islands, and the interior picked up rapidly. Judging from the number of archaeological sites on the Pacific coast, the population increased as fishing intensified. Population densities increased even more after about thirteen hundred years ago, as maritime productivity increased with active upwelling.

The changes are particularly noticeable on the Channel Islands, where food resources were patchy but shellfish and inshore fisheries abounded. After about A.D. 450, sea temperatures cooled for some ten centuries, resulting in strong upwelling and an abundance of fish of all kinds. This was a period when coastal populations exploded and permanent settlements along the Pacific became more commonplace. On the Channel Islands, the major settlements developed more or less equidistant from one another, as if they had distinct, well-marked fishing and foraging territories. Many of these settlements continued in use into historic times, so their inhabitants appear in mission records and in oral traditions recorded in the late nineteenth and early twentieth centuries. We know this because on November 20, 1884, Juan Estevan Pico, a Chumash speaker who lived in the San Buenaventura Indian community at the southern end of the Santa Barbara Channel, compiled a list of no fewer than twenty-one island villages.[11] The names of at least eleven chiefs appear in baptismal records, most of which date to between 1814 and 1822. The rapid depopulation of the islands resulted from the effects of European diseases, the collapse of cross-channel trading, and an extended drought.

On Santa Cruz Island, the largest of the archipelago, most of the larger settlements lay in sheltered bays, where the inhabitants could either command a good view of the coastline or could see nearby villages placed on strategic headlands that served as lookout posts.[12] Each settlement occupied a strategic position close to reliable sources of drinking water. As time went on, territorial boundaries seem to have become more rigid and were often marked by cemeteries. For example, most cemeteries on nearby Santa Rosa Island are associated with major villages, both on the coast and in the interior, some of them perhaps serving as territorial markers.

As populations increased and the islands became more crowded, so fighting escalated. The surge in violence came at a time when signs of dietary stress become more commonplace on the bones of the dead. The biological anthropologists Patricia Lambert and Phillip Walker have chronicled numerous examples of forearm fractures sustained when parrying blows, nasal fractures, healed skull wounds, and injuries caused by spears or arrows.[13] Many of the head injuries healed, having been inflicted by relatively slender blunt instruments.

Both the malnutrition and the increased violence took hold at a time of great climatic instability and marked social change. Santa Barbara Channel waters were generally cool and productive, but on land conditions were generally dry, with serious droughts well recorded in tree rings and deep-sea cores in about 950 and 1250. The droughts resulted in a consolidation of smaller settlements into larger ones, and in the actual abandonment of areas with marginal water supplies around the islands as well as on the mainland.

Violence and increased competition for food were insidious consequences of major drought cycles. On the islands, every settlement was matrilocal—that is, the women stayed in the villages of their birth, while the men married into them. This had a strategic advantage, in that it fostered alliances between neighboring villages where blood-related males dwelled. At the time of the great droughts, we know from linguistic research concerning historic populations that islanders and mainlanders spoke different languages, as if there were two distinct groups. The institution of matrilocal residence may have developed in the face of repeated threats of attack from the mainland, where food supplies were short and territories crowded.

As island populations increased, so the pressure on highly regarded foods like red abalone increased. These meaty shellfish flourish in shallow water and can reach a considerable size. Small groups of women would wade in tidal pools, diving in shallow water to pry abalone off the rocks. They would toss their catch into shallow baskets, swimming across the rocky bottom just below the low-tide level, timing their visits for the lowest tides of the month. Back at camp, they would deftly pry open the abalone with bone levers, and then pound the delicate flesh before roasting it. By the time of the warm centuries, the abalone taken were much smaller, but the food shortfall was overcome by widening the diet to include many more fish species over and above the inshore kelp-bed species favored in earlier times. Just like Great Basin foragers, the Chumash learned the advantages of a broad diet. By 1100, the fishers were venturing into deeper water aboard the sewn-plank canoes known as *tomol*s, taking deep-water species such as shark and tuna with hook and line. They were also using sophisticated toggle harpoons against sea mammals, a technology long in use in the far north. This intensified fishing was a deliberate effort to widen the diet during a harsh drought cycle.

The Channel Islands have relatively few edible plants, and they lack the plentiful acorns so important on the mainland. For centuries, the islanders imported acorns by exchanging them for thousands of shell beads manufactured with fine drills made with the good quality chert found on Santa Cruz Island.[14] At a few island sites, the remains of plants only native to the mainland provide clear evidence of this trade after 1200. Interestingly, the shell bead trade accelerated dramatically during the late first millennium, just at the time when droughts deepened and competition for plant foods intensified. At the same time, the people relied heavily on fish, moving into more permanent villages as they became tethered to local fishing grounds and beaches where they could land and maintain their canoes.

Political and social circumstances on the islands were volatile during the drought centuries. As the population grew at a time of bountiful fisheries, so competition for the best fishing spots and the richest shell beds would have intensified. As we have seen, violence broke out, people were killed, territorial boundaries hardened and were defended more aggressively. Inevitably, too, some individuals or kin groups acquired control of specific fishing areas or desirable plant patches. Such power gave them enhanced authority, and, perhaps, a higher position in new social hierarchies that developed over the generations. Local society became more hierarchical, ruled by local chiefs who formed alliances with one another and sometimes presided over several villages. Some of them were also canoe captains, a specialty so valued that in later times they were members of a formal brotherhood.

On the islands, people responded to drought by living in densely inhabited villages. Their leaders controlled a rapidly growing trade with the mainland that seems to have resulted in a sharp decrease in violence. People suffered from malnutrition on many occasions, but the suffering may have been less than that on the mainland, where drought in the interior decimated acorn crops and drastically reduced stands of wild plant foods. On the mainland, as on the islands, each community adopted the classic response to severe drought: they placed their settlements close to permanent water sources. Judging from the results of crowding in shantytowns on the edge of modern industrial cities, we can speculate that there were serious problems with water pollution and sanitation in now much more crowded settlements, with the inevitable epidemics of dysentery and other diseases that were less common in small villages

In the end, it seems that the island and mainland chiefs developed social mechanisms for damping down the competition and violence that resulted from food shortages. Around 1300, we find complex alliances and other mechanisms for reducing strife coming into play, and these were closely connected with an explosion in island–mainland trade and in exchange with groups living far away in the interior, even as far away as the Southwest. These mechanisms endured long after the great droughts of the warm centuries were forgotten, in a region where unpredictable rainfall and changing sea temperatures made everyone vulnerable to climate change, and where the experience of earlier generations provided precedent for people anchored to coastal fishing grounds or to acorn harvests.

The Chumash maintained contacts not only with neighboring groups, but also with people far inland. Their trade networks extended as far as the Southwest, where Ancestral Pueblo farmers responded to the same catastrophic droughts with different strategies.

A HOT, SULTRY afternoon has settled on the floor of Chaco Canyon, New Mexico. The air is heavy and still, weighing on the visitors passing through the empty rooms of Pueblo Bonito. A huge bank of black thunderclouds masses on the western horizon, mocking the bright sunlight shining on the canyon walls. The clouds pile ever higher. Distant lightning flickers through the menacing overcast; thunder rumbles ominously. Violent wind gusts sweep through the canyon; a few heavy raindrops spatter the sandy path. The visitors run for shelter under a nearby overhang. A wall of gray approaches, but evaporates as the approaching storm changes course and drops its load downstream. Bright sunlight returns to bake the canyon floor once again.

If you live in a semiarid environment, you have to be prepared to roll with the climatic punches. The Ancestral Pueblo of the Southwest were masters of flexible and opportunistic living in landscapes where movement meant survival. A modern-day Tewa Indian oral tradition from the Southwest proclaims: "Movement, clouds, wind and rain are one. Movement must be emulated by the people."[15] Great Basin bands survived through mobility. Coastal fishing groups moved into more densely populated villages. All of these people were hunters and wild plant gatherers, but others cultivated maize and beans and lived by farming the land. Farming was not a viable option in desert areas; California Indians with their acorn harvests had no incentive to grow crops. But, in the Southwest, conservative farmers, who were experts at water management, thrived in often agriculturally marginal environments for over three thousand years.

Chaco Canyon lies in the heart of the San Juan basin, a seemingly empty landscape that appears to stretch forever.[16] Water in any form is short in the basin; summer rainfall arrives as thunderstorms, gentler, if sporadic, winter rains in the form of snow between December and March. But the amount of rainfall is very small, invariably localized, and utterly unpredictable. Short-term climatic shifts such as El Niños or prolonged droughts like those of the Medieval Warm Period could have a profound effect on agriculture on a year-by-year basis. Chaco Canyon is about 18.6 miles (30 kilometers) long and between a third and nearly a mile wide (0.5 and 1.5 kilometers). After storms, the Chaco Wash flows strongly through the canyon, which is also watered by side streams and by natural seeps at the foot of the surrounding cliffs. But in times of drought the canyon floor is bone dry.

Hardly a promising location for any form of agriculture, one would think, but between the ninth and twelfth centuries, at least 2,200 people dwelled within the canyon, and its population swelled considerably at times of major ceremonies. Numerous small hamlets of a few stone houses thrived at Chaco, also nine large pueblos, commonly known as great houses. The permanent residents of Chaco farmed fertile patches of soil with the simplest of digging sticks and hoes. How many of them lived in the great houses rather than in hamlets we don't know, but places like Pueblo Bonito were capable of housing hundreds of people—provided the visitors brought their own food. The carrying capacity of the canyon itself was minuscule relative to the size of the great houses. On the south side of the canyon, the farmers relied on rainfall, while on the north, they cultivated maize and beans in densely packed, gridded fields.[17] By using canals about 4 feet (1.2 meters) wide, they channeled water stored in earthen dams through simple masonry gates from one gridded field to another. The farmers increased agricultural production dramatically as a result, but at a cost of increased vulnerability to severe drought, for their agricultural systems became less diversified. Fine-tuning water control systems on this small a scale would only work if there were enough rainfall to fill the canals and irrigate the field grids. In most years, there was not. The solution was diversification, with each household farming a variety of microenvironments, spreading the risk as one would when investing in mutual funds. But they never produced enough to feed everyone who visited the canyon.

The expansion of Chaco's great houses between 1050 and 1100 depended in large part on a highly vulnerable agricultural system that worked fine in good rainfall cycles, but was disastrous when the rains failed. But, despite this built-in vulnerability, the Chacoans built a series of great houses that were architecture on a grand scale, far larger than anything needed by a mere 2,200 people. Most of the Chaco pueblos were semicircular, often with several stories and numerous rooms associated with large and small kivas, subterranean chambers used for rituals of all kinds. The great houses are so unexpected and spectacular that some people call Chaco the Stonehenge of America.

No one knows why such imposing great houses rose in this remote, arid canyon, but there is universal agreement that Chaco became a place of supreme ritual importance between the ninth and twelfth centuries. People traveled here from far away, bringing grain, timber beams, turquoise for ornaments, and exotic commodities such as tropical bird feathers, presumably at times of major ceremonies such as those that marked the passage of the solstices. The most imposing of the great houses, Pueblo Bonito, was begun in about A.D. 860 and enjoyed a complex architectural history that lasted until about 1115. By this time, Chaco exercised a profound cultural and spiritual influence over an enormous area of the San Juan basin. A series of symbolic and irregular "roads" radiated from the canyon, but their function and symbolic significance eludes us.[18]

Experts have debated the significance of Chaco for generations. Perhaps the most powerful individuals in the canyon were religious leaders. They were guardians of the spiritual knowledge that governed a society where religion and agriculture went hand in hand. By combining agriculture and clever water management with complex ritual beliefs, the people flourished in times when rainfall was relatively abundant. Here, human existence depended not only on water management, but, they believed, on the meticulous performance of elaborate rituals and dances, on the skill of people who were guardians of sacred knowledge. This was why such stupendous great houses rose in an arid canyon, where the setting was spectacular. People came from many miles away, drawn by the powerful spiritual associations of the place.

Chaco had developed rapidly during a long period of relatively plentiful rainfall by San Juan standards. Tree rings tell us that the persistent and increasingly severe droughts that affected so much of the American West descended on the canyon after 1100, at a time when Chaco's need for timber and building labor was at its height. Agricultural productivity

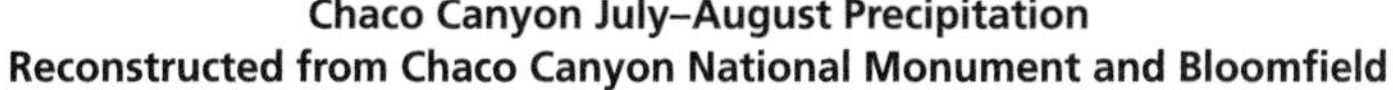

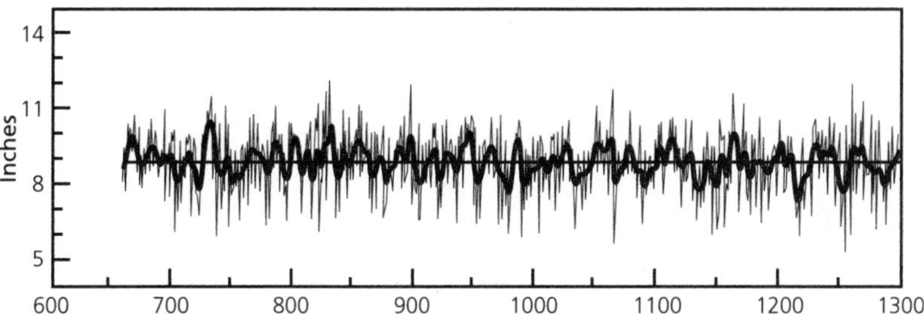

A plot of rainfall in July–August at Chaco Canyon, New Mexico, A.D. *600–1300. Reconstructed from tree-ring sequences from the canyon using Bloomfield's method for analyzing time series. The statistically derived dark line shows general trends, while the spiky lines denote exceptional rainfall years. The drought of the 1140s and the thirteenth-century great drought can be clearly discerned.*

withered; water supplies slowly evaporated. To the people who lived stoically through the drought, it must have seemed as if the forces of the supernatural had deserted them, that their leaders had lost their ability to communicate with the gods. Within a half century, the great houses were deserted. The Chacoans had moved elsewhere.

THIS PATTERN OF movement was a familiar strategy in an area where rainfall ebbed and flowed and no single community was completely self-sufficient. No one at Chaco, or elsewhere in the Southwest, had any illusions about crop failure and hunger. Every family maintained kin ties with communities elsewhere, traded with them, responded with food when need be. By the same token, people living in better-watered areas knew that their kin would move to live among them if conditions for farming were better than at home. Chaco itself exercised an influence over a wide area, especially to the north and west, where people moved after 1100. Other pueblos rose to prominence, among them Aztec above the Animas River; then, even farther north, great communities flourished in the Moctezuma Valley and in Mesa Verde, places where densely populated settlements lay close to more dependable water supplies.[19] But, once again, a severe drought from 1276 to 1299 caused people in these great pueblos to disperse to escape conflict as more people crowded into great houses and food supplies grew increasingly scarce. High infant mortality rates and reduced birthrates owing to health problems may have caused local population densities to fall, but also the social and technological means to deal with the crisis were inadequate. So the people relied on their ancient strategy of movement, going to live in better-watered areas to the east, south, and west, where there were already communities with friendly kin capable of supporting large numbers of people.

For the Pueblo, movement was the only logical solution to the droughts of the warmer centuries. We know there was hunger and fighting for food and water. There is even evidence of ritual cannibalism. Fortunately, the ancient traditions kicked in and people adjusted by moving away, household by household. In some pueblos, large numbers of clay pots and stone tools, also heavy grinding stones, still lie where their owners left them. The strategy of movement was successful. Centuries later, the legacy of the Ancestral Pueblo thrives in vibrant oral traditions and ritual observances, in architectural and agricultural practices that resonate in the Southwest to this day. Despite all the trauma of the Spanish *entrada* and the upheavals wrought by industrial society, Pueblo communities are politically autonomous today, using ancient shared rituals to promote common identity and harmony. Chaco Canyon and other pueblos remain part of the ancestral memory of modern Pueblo societies, and visitors still go to breathe in the strength of the ancient places and their inhabitants.

There was no mass migration out of the Chaco, no one moment when the elders abandoned the great houses that were the focus of life in the canyon. The close deliberations took place in subterranean kivas around winter fires, also at home between husbands and wives as their hungry children slept. Perhaps several husbands would trek out from the canyon to visit relatives, to prepare the way for a move to somewhere where they would be welcome and there would be enough food to tide them over until next cycle of planting and harvest. Then two or three families would pack up their possessions, abandoning the small homestead where their ancestors had dwelled before them. They left with regret, perhaps, but comfortable with their decision, as they knew that the ancestors would have approved of it.

The great droughts of the American West had their origins in complex and still little-understood interactions between ocean and atmosphere in the Pacific. These same interactions, reflected in the cool cycle of the Pacific Decadal Oscillation and in the continual, unpredictable swings of El Niños and La Niñas, also brought havoc and food shortages to elaborate civilizations in Central America and the Andes.

CHAPTER 8

Lords of the Water Mountains

Again there comes a humiliation, destruction, and demolition. The manikins, woodcarvings were killed when the Heart of Sky devised a flood for them. A great flood was made: it came down on the heads of the manikins, woodcarvings.

—*Popul Vuh*, the Maya book of dawn of life[1]

HUNTING BANDS IN NORTH AMERICA'S Great Basin stayed close to shrinking water supplies and moved to higher ground. The Ancestral Pueblo abandoned Pueblo Bonito and the other great houses in Chaco Canyon, for centuries the sacred hub of their world, in the face of intense drought. Chumash Indian groups along the southern California coast fought each other over water and precious acorn crops devastated by persistent drought. All these peoples lived in semiarid landscapes, where flexibility and movement were automatic reactions to drought. Deep traditions of mutual obligation, of reciprocity, were a fundamental part of life in the American West, where the endlessly cycling pumps of deserts sucked in and pushed out people as rainfall came and went. Far to the south, in Central America, ancient Maya civilization was at

Map of the Maya lowlands showing locations mentioned in the text.

its apogee when the droughts of the warm centuries arrived, bringing catastrophic disruption that killed thousands and depopulated much of the southern Maya lowlands.

JUST AS, IN the 1970s, the decipherment of their script revolutionized our knowledge of the ancient Maya, so in the last few decades have deep-sea and lake borings provided insights into the climatic changes that rippled across their rainforest homeland a thousand years ago. The evidence for repeated, and often severe, drought mirrors that from western North America

The Cariaco basin, off Venezuela in the southeastern Caribbean, is the source of the most influential of all deep-sea cores, because it records climatic shifts resulting from the northward and southward movement of the Intertropical Convergence Zone (see sidebar in chapter 4). The ITCZ had a profound effect on rainfall in the Maya lowlands and across the ocean, on the Saharan Sahel. We can, in general terms, link climatic changes on either side of the Atlantic. As we shall see in chapter 11, there may even be connections across the Pacific to Asia. The Cariaco core is remarkable for its layers of fine sediment deposited annually by rivers flowing into the ocean.[2] The deposits are unusual, for they are exceptionally well defined, with about 11.8 inches (30 centimeters) representing every millennium. The laminated sediments reflect fluctuations in river output caused by changing rainfall amounts. The changes result from seasonal shifts in the Intertropical Convergence Zone, dark laminae reflecting rainfall during the summer and fall rainy season, light-colored ones in the dry winter and spring, when the Convergence Zone is at its southernmost position and trade winds blow strongly along the Venezuelan coast. The bulk titanium content in the laminae records the amount of land-based sediment brought to the basin from surrounding watersheds. The higher the titanium content, the greater the rainfall.

Among sediments deposited over the past two thousand years, titanium concentrations were lowest between about five hundred and two hundred years ago, during the dry centuries of the Little Ice Age. Higher titanium concentrations arrived in the basin between A.D. 880 and 1100, the heart of the Medieval Warm Period. But the titanium levels were far from constant. There were pronounced minima in about A.D. 200, 300, and 750. Fortunately for climatologists, the Cariaco basin lies within the same climatic regimen as the Maya lowlands, where most rain falls during the summer, when the Intertropical Convergence Zone is at its north point over the Yucatán. If the Convergence Zone remains at its southernmost position for any length of time, drought hits both the Cariaco region and the Maya homeland.

Maya civilization flourished before about A.D. 150, when cities like El Mirador reached enormous size. El Mirador was abandoned rapidly during the early first millennium A.D., at a time when the Cariaco sea cores tell of drought in the area. But the Maya recovered, new cities arose, and new water management strategies came into use. Relatively wet times between A.D. 550 and 750 saw growing populations. Many Maya communities were soon operating at the limits of the carrying capacities of their lands. By this time, Maya settlements large and small were far more vulnerable to multiyear droughts, which would descend without warning but so infrequently that they lay outside the short generational memory of the day.

The Cariaco core documents a low-titanium interval centered on the ninth century, at the beginning of the Medieval Warm Period, a severe drought also found in a core bored in Lake Chichancanab in the Yucatán itself.[3] The two records chronicle multiyear droughts that began as early as A.D. 760, and reoccurred at about fifty-year intervals: 760, 820, 860, and 910. The first drought was a slight long-term drying trend, followed by a severe drought of about three years beginning in 810, and another drought beginning in about 910 and lasting for about six years. It was during this period that Maya civilization in the southern and central Yucatán lowlands collapsed.

The Lake Chichancanab core not only mirrors that from Cariaco, though it can be dated somewhat less accurately (plus or minus twenty years), but also documents drought conditions lasting until A.D. 1075. The climate record is now unequivocal. Drought cycles during the early Medieval Warm Period settled over the Maya lowlands at about fifty-year intervals, at the same time as profound aridity affected western North America.

THE MAYA WERE obsessed with water, and with good reason, for they lived in an environment of uncertain rainfall. They believed that civilization began in the dark primordial waters of Xibalba, the Otherworld. The waters were calm and dark, the world nothing but water. "There is not yet one person, one animal, bird, fish, crab, tree, rock, hollow canyon, meadow, forest. Only the sky alone is there; the face of the earth is not clear. Only the sea alone is pooled all under the sky; there is nothing whatever gathered together. It is all at rest." Xibalba was a calm, dark pool. But there were signs of movement in the water, "murmurs, ripples, in the dark, in the night." The creators were in the waters, "a glittering light. They are there, they are enclosed in quetzal feathers, in blue-green."[4] Here the gods created humanity. Once they had done so, they caused water to well out of the entrances to the nether regions, the most sacred places in the Maya landscape, to nourish crops and nurture human life. Water defined this most flamboyant of native American civilizations in subtle ways, and became an instrument of political power and social control. The cycle of Maya life ended when great lords descended in death to Xibalba's inky waters. And when the rains failed and drought descended on the Maya world, the foundations of civilization quivered.

Like the ancient Egyptians, the Maya were village farmers for many centuries before they transformed their homeland into a landscape of great cities ruled by powerful lords.[5] There the resemblance ends. No annual inundation or fertile river floodplain provided a safety net for Maya civilization. Nor did any great rivers link cities, towns, and villages into a large, unified state like that of the pharaohs. The Maya farmed the Petén-Yucatán peninsula, which juts into the Gulf of Mexico; it is a vast limestone shelf lifted from the depths of the ocean over an immense length of time. They lived in a densely forested world where huge mahogany trees towered as much as 150 feet (45 meters) above the ground, where sapodilla and breadnut abounded. The forest gave way to patches of open savanna covered with coarse grass and stunted trees. Hot, humid, and generally poorly drained, the Maya lowlands were a fragile, water-stressed environment even at the best of times. The porous limestone bedrock absorbed water to the point where fluctuations in the water table were unpredictable. It's hard to imagine a less likely place for a great civilization.

Fly over the homeland of Classic Maya civilization (A.D. 250–900) in the lower Yucatán and you pass over a featureless green carpet. But the seeming uniformity is an illusion. The dense tree cover masks an astonishing diversity of local habitats, all of which presented special challenges to ancient Maya farmers. Between 53 and 78 inches (1,350 and 2,000 millimeters) of rain fell each year, but precipitation was less abundant and predictable than one might expect. Most rain arrived between May and October, after a dry season that lasted between four and six months. Lakes, springs, and perennial streams were priceless rarities. Even small-scale village farming required creative ways of collecting, storing, and managing water for the long dry months.

To UNDERSTAND WHY, we must journey back to the beginnings of the Maya. In about 1000 B.C., only the coastal plain and a few small, perennial drainages could support permanent farming communities for the long term. Many of these communities were also involved in fishing. As local populations increased, so small groups moved inland along streams and the swamps that bordered them. Within six centuries, a patchwork of tiny villages had become a rapidly growing civilization.

By the third century B.C., thousands of people lived in scattered communities across the lowlands. They were adept, patient farmers, who developed many ways of managing water and modifying the environment for maximum productivity. Generations of trial and error, of hunger and plenty, led people to settle at strategic locations—for example, they could live near the base of shallow natural depressions that received surface runoff during the rainy season.[6] There they constructed reservoirs to store rainwater, close to places where the first great ceremonial centers rose with remarkable speed. By any standards, these were imposing structures. For instance, between 150 B.C. and A.D. 50, a mere two centuries, El Mirador in Guatemala's Petén grew to cover 6 square miles (16 square kilometers), dominated by the Danta Pyramid, built on a natural hill more than 230 feet (70 meters) high.[7] The city lay amidst an undulating landscape, where water collected in a large, basically natural watershed during the rainy season. Causeways traversed low-lying swamps or shallow lakes near the central area. El Mirador thrived thanks to a simple form of water management, by which large natural depressions, extended into reservoirs, served as water storage facilities. The same reservoir system was also an integral part of an extensive agricultural landscape.

Some of these systems reached considerable size. The people of Edzná in Campeche, occupied between 400 B.C. and A.D. 150, excavated huge canal basins and removed nearly 62 million cubic feet (1.75 million cubic meters) of fill to build them—more than the volume of Teotihuacán's vast Pyramid of the Sun far away in the Mexican highlands.[8] Then, suddenly, during the first and second centuries, El Mirador, Edzná, and the other growing centers collapsed. Their pyramids and temples were abandoned to the forest; the people dispersed into small villages scattered across the landscape. Many experts believe this was because of a severe drought that rendered their reservoirs and simple water systems virtually useless.

After A.D. 250, the start of the Classic period of Maya civilization, water management strategies changed. For the first time, Maya lords built palaces, pyramids, and temples on elevated ridges and hillocks. They moved their centers away from natural water sources and chose to build large reservoirs close to their ceremonial precincts. The largest cities, among them Calakmul, Copán, and Tikal, developed elaborate water management systems, built as part of the process of erecting the plazas and pyramids that were symbolic replicas of the Maya world. The pyramids became "water mountains." Thousands of villagers labored over the great centers. The first order of business was to quarry stone, to create the artificial depressions that would become reservoirs and tanks. Without a viable water system, the builders had no drinking supplies for work gangs or for mixing the limestone mortar used for floors and walls. The labor investment was enormous, but no urban center could survive without far larger-scale water supplies than those required by a small farming village.

Every Maya lord built his largest reservoirs and tanks close to his most imposing civic architecture. Here he appeared before the assembled populace in elaborate public ceremonies. Here dancers would perform in open plazas dwarfed by high temples. Here chants would resonate from pyramid stairways, torches flickering and flaring in the night breeze as incense wafted among the crowd. The ruler himself would emerge from a dark opening in the temple, the symbolic entryway to Xibalba, and appear before the people in a state of trance induced by hallucinogenic drugs. He performed ritual bloodletting on himself, then vanished abruptly, embarking on a journey into the supernatural world. The great rituals revolved around the complex relationships between the living, the deities, and the ancestors, between the rulers and the ruled. The same ceremonies recognized the central place of water in Maya life—and in the complex equations of political power. From earliest times, Maya society had thrived on kin ties that linked community to community with vital ties of reciprocity—the obligation to provide food, assistance, or labor to kin in times of need. As Maya civilization emerged from its village roots and new leaders came into prominence, ancient kin ties still provided at least a theoretical link between lords and commoners. But the nature of the ties between them changed from simple reciprocity to a more elaborate social contract. Maya rulers proclaimed themselves divine lords, with exceptional supernatural powers. They became like shamans, capable of passing freely into the supernatural world, where they served as intermediaries between the living Maya, their ancestors, and the forces of the spiritual world. The lords provided supernatural protection, fostering rain and good crops. In exchange, the people paid tribute and taxes in food and labor. They worked when called upon for the nobility, for the public good, in an unwritten alliance that justified social inequality. As long as there was ample rainfall, the tacit contract survived. But once the fallibility of the lords became apparent, social disorder ensued.

TIKAL, THE GREATEST of the Maya cities, was dominated by the most imposing of all water mountains, founded during the first century A.D.[9] Thirty-one rulers, the earliest dating to A.D. 292, the latest to A.D. 869, left nearly 600 years of recorded history here. By a combination of judicious political marriages and warfare, Tikal's lords extended their sway over neighboring centers, ruling at one point over between 200,000 and 300,000 people. No one knows how many people lived in the center, surrounded as it was by a hinterland of villages and larger communities.

But we do know that the city depended entirely on seasonal rainfall for its water supplies.

The amount of water collected at Tikal was truly astounding. Six catchment areas surrounded the major hillock where the city lay. One central-precinct catchment alone covered 156 acres (63 hectares) and could collect more than 31.7 million cubic feet (900,000 cubic meters) of water in a year when 59 inches (1500 millimeters) of rain fell—an amount typical of a nondrought period. Slightly canted pavements and subtly diverted weirs directed rainfall into the central-precinct reservoirs, which were sealed with stones and imported clay. The combined reservoirs could hold between 3.5 million and 7 million cubic feet (100,000 and 200,000 cubic meters) of water, enough to allow controlled release of water during the dry season, using carefully placed sluice gates under the city's causeways. Small domestic tanks for the houses immediately below the summit were probably recharged from the central system. Four large reservoirs lay near the foot of the central hillock and around the swampy margins of Tikal, designed to recapture gray and reused water from upslope residential areas. The 1.76 to 6.1 million cubic feet (50,000 to 175,000 cubic meters) of runoff served to irrigate about 210 acres (85 hectares) of swamp-margin plots around the city.[10] With water available year-round, local farmers could potentially raise two crops a year.

Tikal's water system was huge and complex, a startling contrast to the much less elaborate water systems that sustained small communities. Most villages survived the dry season by using shallow water tanks close to the settlement that held enough water to carry the farmers through one dry season, perhaps a few months more. But Tikal's vast reservoirs held enough water to reduce the vulnerability of its residents in dramatic ways. Two or three years of little or no rain would not cause problems for Tikal's water managers, though such a long drought would have drastic consequences for the small villages that held most of the dispersed Maya population. But even Tikal's water system was inadequate for sustained, multiyear droughts.

THE HUSHED CROWD in the plaza gazes upward to the temple at the summit of the pyramid. Flickering torches cast deep shadows in the gloom of dawn. Incense smoke drifts across the slopes of the sacred mountain. High above the mob, serried rows of white-clad nobles surround the dark entrance to the shrine. Suddenly, the great lord appears, his hair long, tied above his head with brightly colored feathers that cascade down his back. He is bare-chested, wearing a brilliantly white loincloth, his legs and wrists adorned with deep, blue-green jade beads. A noble in a white cape lays a broad clay bowl holding unmarked paper and a stingray spine before him. The lord squats, pierces the loose flesh of his penis three times, and threads paper strips through the wounds. The tan paper turns bright red as the lord dances himself into a frenzied trance, holding a symbol of the double-headed serpent that symbolizes the path of communication with the gods. Conch shell trumpets sound; a god has been summoned from the Otherworld. The crowds milling in the plaza sway in ecstasy as the drums beat. Brightly dressed nobles dance on a terrace below the lord. The devout gash themselves and spill blood onto cloth bands on their arms and legs.[11]

Great Maya lords proclaimed themselves divine rulers, related by carefully crafted genealogies to prominent ancestors and the gods themselves.[12] Kingship and the world that defined Maya civilization were closely tied to the experience of the humble villager. Rulers molded their power, and the symbols of that power, from the plants and animals of the forest, from the ancient rhythms of planting and harvest, from the alternation of the dry and rainy seasons. Like their subjects, they looked at the world in the context of things both spiritual and human, ancestral and contemporary, and in the context of realms that were those of lords or of commoners. The bloodletting, the trances, the elaborate ceremonies that surrounded the accession and death of kings, were an integral part of a society that gave lords the right to control the water supplies that came from heaven. And water mountains were the conduits that brought water from the spiritual realm to the earth.

The Tikal water mountain with its reservoirs and tanks allowed its rulers to control water supplies for large numbers of people at lower elevations. Maya lords promoted themselves as divine leaders with powerful supernatural abilities. But their real power came from their control of key resources such as water, and from the reality that many of their subjects lived in engineered landscapes. Maya life revolved around the seasons of planting, growth, and harvest, each with its ceremonial associations, so water rituals were an essential part of the fabric of daily Maya existence. At Tikal and other cities, the lords manipulated water not through authoritarian rule, but by using ritual to direct and appropriate the labor needed to excavate and operate the system. It was no coincidence that their water mountains were settings for elaborate public ceremonies.

We know little of these rituals, but we do know that two metaphors defined death for the Maya nobility. One was a fall into a watery underworld, often into the open jaw of an earth monster—a chasm in the surface of the earth. Another was a journey by canoe into the endless waters below the earth. Both metaphors linked lords to water, to the mirrorlike surfaces of the teeming reservoirs that lay close to their pyramid burial places. Whether one lived near a major center or in a remote village with its own water tanks, survival ultimately depended on the sustainability of the large cities—the water mountains. During periods of prolonged drought, farmers from outlying areas would migrate to the edges of the great cities, to the feet of the artificial mountains that were the anchors of Maya water supplies.

As Maya lords developed their enormous reservoirs, so the predictability of water supplies increased—up to a point. But the only source of their water was rainfall. Unlike societies that could draw a relatively constant supply of water from rivers or subterranean aquifers, the Maya, for all their engineering prowess, were highly vulnerable to short-term discontinuities of climate.

They lived in an environment of constant disruptions—years of drought and crop failure, torrential rains and soil erosion, unexpected storms that drowned their crops. They farmed with the simplest of methods, but with a comprehensive knowledge of their forest environment. Like other tropical farmers, they used slash-and-burn methods, cutting down a patch of forest, burning the wood and brush, then working the natural fertilizer of ash and charcoal into the soil. They would plant with the first rains, use the garden for two years, then abandon it, as the soil lost its fertility rapidly. These milpa plots were a patchwork of newly cleared and regenerating land, surrounded by thick forest that vanished progressively in the face of rising populations over the centuries. Cultivating these lands required great experience and infinite patience, for pelting rain and intense tropical sunlight soon hardened the soil. But the lowlands are far from uniform. Fortunately, the Maya lived in a diverse environment, where they could also practice other forms of agriculture. In swampy areas, they built raised-field systems, with narrow rectangular plots raised above swamps or seasonally inundated lands bordering rivers. These productive fields could yield several crops of beans and maize a year. The farmers also terraced steep hillsides, the stone-faced terraces trapping silt that would cascade downslope during torrential downpours.

Whatever farming methods they used, the Maya were brilliant agriculturalists. They grew a wide variety of crops, suited to a wide diversity of microenvironments, and selected arable land with the utmost care. Everything was a mosaic—of highly productive swamps and raised fields, milpa plots, and terraced hillsides. The Maya managed and manipulated their environment over the centuries, always living in dispersed communities, even close to major centers, for the realities of their homeland would never support large concentrations of people living in one spot. But the population was denser than it appeared. In parts of the southern lowlands, population densities rose as high as 600 per square mile (2.6 square kilometers) over an area so large that people could not move away from their local environments if drought or other disasters came along. The landscape filled up. As urban populations rose, so the Maya ate up their land, at a time when the farmers were supporting a growing nobility and increasing numbers of nonfarmers.

Maya civilization was never a centralized state like that of the pharaohs or the Babylonians. The decipherment of Maya glyphs during the 1970s, one of the great scientific triumphs of the twentieth century, revealed a landscape of frantically competing city-states governed by ambitious and rapacious lords obsessed with genealogy, warfare, and personal advancement. Tikal, for example, rose to prominence in the first century B.C. By A.D. 219, the lord Xac-Moch-Xoc had founded a brilliant ruling dynasty that conquered its nearest neighbor, Uaxactún. Three centuries later, the dynasty presided over a territory of 965 square miles (2,500 square kilometers). The city was but one of many centers vying for power in a volatile, ever-changing world. Alliances were forged and would then fall apart as a lordly partner died. Rulers would conquer their neighbors, sacrifice their leaders, and cement a new relationship with a timely diplomatic marriage. But, ultimately, this entire landscape of political wheeling and dealing, of warfare, elaborate ritual, and kingship, depended on water from the heavens. When the now well-documented Medieval Warm Period droughts descended on the lowlands, they eroded the foundation of Maya civilization.

WERE THE DROUGHTS enough to cause the collapse of Maya civilization?[13] Each planting season, the farmers gambled with their crops, setting seed in the ground with the first rains, then waiting for later storms to moisten the drying soil. Some years the rain would fall. In others, weeks would pass with mounting clouds on the horizon, but no showers except for a few heavy drops. Rain would fall a few miles away, but the dark cloud would bypass other villages. In hundreds of small hamlets and tiny farming communities, people lived from harvest to harvest, just as they did in distant medieval Europe. Everyone experienced episodes of hunger during their lifetimes. To what extent the lords redistributed food to hungry communities as part of their rituals, we do not know.

In the early centuries of Maya civilization, there was a natural cushion of wild foods that one could rely upon in dry years. But rural populations rose inexorably, and in many places stressed the carrying capacity of the forest; now people ate up the land as they cleared more brush and trees, exposing fragile soils to the harsh sun. The primordial forest vanished, replaced by regenerated growth. Fewer and fewer wild resources remained as famine foods in dry years. The margin between plenty and hunger, between good harvest and poor, narrowed significantly between the seventh and ninth centuries.

The population growth came during generations when the number of nonfarmers increased, when more and more people aspired to the nobility. Increasing numbers of high officials, traders, and priests now claimed ancestry from noble lineages. Maya society became top-heavy with nobility, with non–food producers, with people keenly aware of the privileges that came with rank.[14] The demands on commoners, on rural farmers, in terms of food and tribute, increased steadily from generation to generation. As the rural population ate up the land, deforestation accelerated and agricultural yields reached their limits. Meanwhile, the elite were largely divorced from the harsh realities of aridity and hunger that would descend on their homeland during multiyear droughts.

Even at the distance of more than a millennium, we can see the demise of Maya civilization in the southern lowlands playing out like a Greek tragedy. The droughts begin in the early ninth century, at the beginning of the Medieval Warm Period. After a year or two, villagers are hungry, but still subject to inexorable demands for food and tribute. The water mountains, for example Tikal and another great city, Copán, with its nearby river, still have adequate water reserves, but these reserves are diminishing. The public ceremonies linking great rulers to the smooth mirror of primordial waters still unfold. In this sense, the lords are active participants in the shaping of history, for they bring their own historical consciousness to bear and respond to drought as they always have, by distributing food and appeasing the forces of evil, by ritual, strategic alliances with former enemies, and by warfare. But the drought continues: the reservoirs begin to run dry.

For generations, the people have considered their lords, the descendants of divine ancestors, to be infallible guardians of the harvest, of Maya life. But now they have feet of clay, are powerless in the face of mocking, cloudless skies, and unrelenting heat. Tikal and Copán and the other cities like them fall apart.[15] Social disorder erupts in the face of persistent hunger and water shortages. The commoners rise in protest against the bloated nobility; they desert their leaders and scatter through the countryside, leaving mere handfuls of survivors, some the descendants of great lords, still squatting among the ruins.

An apocalyptic scenario, perhaps, but entirely plausible, given the vulnerability of Maya civilization to multiyear droughts, of the kind now known from climatological records. The droughts themselves did not destroy Maya civilization, but the economic, political, and social consequences of arid years certainly did. By the late tenth century, the great cities of the Petén and southern Yucatán had imploded as their inhabitants scattered in the face of hunger and chronic water shortages. As had happened thousands of years before in Mesopotamia and along the Nile, drought and famine brought social unrest, rebellion, and the collapse of rigid social orders that were based on doctrines of lordly infallibility.

Of course the fall of Maya civilization was more complicated than this, for intricate political and social factors came into play. In some places, the elite continued to wage war as they had always done, even at a time when civilization was collapsing around them. The archaeologist Arthur Demarest has spent five years excavating six Maya sites, among them a heavily fortified center, Dos Pilas in the Petexbatun area of northern Guatemala.[16] He believes that the raiding form of warfare favored by Maya lords in earlier times gave way to all-out civil war in this area during the eighth century. The center's rulers became ever more aggressive and conquered their neighbors. Eventually the kingdom became so large that it fragmented into smaller warring chiefdoms, each with its own fortified center. Inevitably, this activity and endemic warfare had adverse effects on the delicate jungle ecology. Dos Pilas has miles of trenches and moats. Demarest's excavators have found numerous spearheads at the foot of the walls, caches of decapitated skulls, and postholes that mark now vanished palisades and towers. He believes that the effects of all-out war were disastrous, as the violence forced changes in farming methods, causing soil depletion and then widespread crop failure. Concentrated agriculture in strategic locales simply did not work in the Maya environment, especially at a time when multiyear droughts were stressing the already overtaxed water systems. In the end, a nobility obsessed with warfare hastened the demise of one of the Americas' most dynamic and innovative civilizations.

After about 1100, climatic conditions became more humid, but Maya civilization in the southern lowlands never recovered. In the northern Yucatán, large cities and centers continued to flourish, thanks, in large part, to natural sinkholes, *cenotes*, in the limestone that allowed people to reach the water table. Maya civilization endured, albeit on a reduced scale, until Hernan Cortés and his Spanish conquistadors arrived over the eastern horizon in 1519. But what would have happened if more cyclical, multiyear droughts had afflicted the Maya lowlands? What if warming had continued and the water table had dropped so that the *cenotes* dried up? Inevitably, there would have been another implosion, this time of cities and kingdoms in the northern parts of the Maya world, where people, once again, would have adopted the only defense available to them: dispersal into small, largely self-sufficient communities. Would, then, have Cortés and his motley band of adventurers ever marched to the heart of Aztec civilization in the highlands? In a different, semiarid environment, would Aztec civilization ever have come into being, founded as it was on conquest, tribute, and an agricultural base of swamp gardens that depended on the swampy lands of the Basin of Mexico? The historical possibilities are intriguing.

The implosion of lowland Maya civilization is a sobering reminder of what can happen when human societies subsist off unpredictable water sources, and, through their efforts, put more demands on the water supply than it can sustain. They may build water mountains or hundreds of acres of irrigation canals, but, in the final analysis, they are powerless against the forces of drought, flood, and El Niños, especially when their rulers are oblivious or indifferent to the suffering of those who feed them. The analogies to modern-day California, with its aqueducts for water-hungry Los Angeles, or to cities such as Tucson, Arizona, with its shrinking aquifers and falling water table, are irresistible.

Below the equator, the warm centuries and their epochal droughts also brought powerful climatic shocks.

CHAPTER 9

The Lords of Chimor

The bodies of the kings and lords were venerated by the people as a whole, and not just by their descendants, because they were convinced . . . that in heaven their souls played a great role in helping the people and looking after their needs.

—Father Bernabé Cobo, *Historia general de las Indias* (1653)[1]

MOCHE VALLEY, COASTAL PERU, A.D. 1200. The gloom hovers low over the cliffs, the Pacific calm and oily smooth in the early morning. Fog has lingered for weeks, ebbing and flowing with the daylight. Just the regular beat of the ocean surf resonates over the low-lying shore, where the reed-matting shelters of the fisherfolk huddle close to the sand. Campfires flicker in the gray; cloaked figures move in and out of the shadows. Piles of reed canoes lie along the beach, well clear of high tide, drying after a few days of use. At the edge of the breakers, young men launch one of the canoes through the crashing swell, leaping nimbly aboard as the curved bows rise to a steep wave. They laugh as they paddle close offshore, watching the seabirds soar and weave overhead. Soon they bear to the left, toward a spot where gulls dive steeply into rippling water alive with hundreds of small fish. The birds shriek and move elsewhere as the men cast light fiber nets and baskets into the teeming mass of anchovies. Quickly, they scoop the hungry fish aboard until their laden canoe floats deep in the water. Then they paddle for the beach, shooting the waves with effortless bravado, landing smoothly on the sand. Waiting men and women lift the reed boat and carry it above the high-tide line. Quickly, they offload the silvery cargo. The young skipper immediately heads out for another load, while the people ashore begin the process of drying and grinding the anchovies into meal.

Sixty miles (100 kilometers) inshore, fish meal from the coast makes its way inland, wrapped in cloth bundles that are carefully balanced on the backs of a llama caravan. The animals follow a well-maintained road along the fertile river valley, where patches of green maize in carefully tended fields stand out against the buff-colored desert. The llama have traveled for days, from high in the Andes to the coast, bringing oca, potatoes, ullucu, and other carbohydrate-rich foods from the mountains.[2] Now they are on the return journey with fish meal and dried seaweed, the latter used to combat the goiter that plagues highland farmers with no source of iodine for their diet. Fish meal, cotton, and seaweed are staples of a trade that has flourished as long as anyone can remember.

The arid north coast of what is now Peru was one pole of an interconnected highland and lowland Andean world, where long-term droughts and El Niños were a constant threat to civilization.

Teleconnections, links between contemporary climatic events in different parts of the world, are fundamental to understanding the Medieval Warm Period. Climatologists tie together such events as droughts or major El Niños by using ice cores, tree rings, deep-sea cores, and other well-dated proxies, but even the best-documented links remain tentative. However, for the first time, we are beginning to learn just how widespread droughts were during the warm centuries. A grid of tree-ring sequences reveals prolonged dry cycles in the North American West and Southwest between the tenth and thirteenth centuries that appear

Locations mentioned in this chapter. Some obscure places are omitted.

to be linked to the Pacific Decadal Oscillation and to cool conditions in the eastern Pacific. The Cariaco basin core from the Caribbean and borings from Yucatán lakes document harsh droughts in the Maya lowlands of Central America between the eighth and eleventh centuries that can be linked to the Intertropical Convergence Zone lingering in a more southerly position. Can we, then, add Andean droughts to this emerging pattern of dry cycles? If so, can we reconstruct how the wealthy civilizations of the coastal Peruvian desert adapted to lengthy droughts?

The eighteenth-century German naturalist Alexander von Humboldt was the first European scientist to explore the Andes. He marveled at the great variety of animals and plants that thrived in the harsh and varied landscape of the high peaks. Humboldt observed the constant assault on higher altitudes waged by mountain farmers, who lived among terraced fields on the sides of deep valleys. The mountains tumble through steep foothills to the low-lying Peruvian coast, a shelf at the foot of the Andes crossed by rivers that descend from the mountains to the Pacific. The bleak coastal plain is itself effectively rainless. It is carved by some forty rivers and streams fed by mountain runoff, but much of the low-lying terrain is too sharply undulating to permit irrigation. Two large areas on the north coast boasted of topography that allowed farmers to link their field systems to canals: the Chicama-Moche valleys and the Motupe-Lambayeque-Jequetepeque region.

Few human communities anywhere, let alone civilizations, have ever faced such a challenging environment—not only one of the driest on earth, but one where short- and long-term climatic events could alter the landscape drastically almost overnight. In good years, there was enough water for farmers to cultivate maize, beans, cotton, and other staples, provided they managed the runoff carefully and irrigated their fields. For agriculture, everything depended on canals, reservoirs, and expert water conservation strategies. Fortunately for the farmers, the people of the coast did not live on agricultural produce alone. The Pacific was a bountiful source of shellfish, and, above all, of anchovies, which abounded by the million close inshore thanks to the natural upwelling of nutrients stimulated by the cold, north-flowing current (now named for Von Humboldt). Before industrial exploitation depleted the fishery, the swarms of anchovies were almost unbelievable. In 1865, the American archaeologist and diplomat Ephraim Squier described rowing in a small boat through the sluggish swell, passing through "an almost solid mass of the little fishes which were apparently driven ashore by large and voracious enemies in the sea." A 1.2-mile-long (2-kilometer) belt of fish lay so close to shore that women and children were scooping them up "with their hats, with basins, baskets, and the fronts of their petticoats."[3]

Against seemingly insurmountable odds, a series of wealthy states flourished along Peru's north coast for many centuries. A combination of agriculture, fishing, and long-distance trade provided at least a partially successful buffer against an arid, demanding environment.

In good rainfall years, agriculture in coastal river valleys could be highly productive, provided water was carefully managed. Even in good times, the people of each town and village conserved every drop they could, for they well knew that even a one- or two-year drought could bring hunger. But ice cores tell us that the Andeans faced far longer dry cycles than that. Lonnie Thompson of Ohio State University has long worked on ice cores from the Quelccaya ice cap in the southern Peruvian Andes. His sequences reveal four widespread regional droughts between the sixth and early fourteenth centuries A.D., culminating in a major dry spell between 1245 and 1310.[4] This catastrophic thirteenth-century drought had drastic effects on highland societies, especially the pre-Inca state of Tiwanaku, centered along the southern shores of Lake Titicaca.[5] Lake core data from Titicaca itself date the onset of the drought here to about 1150, somewhat earlier than elsewhere. The lake level dropped between 39 and 56 feet (12 and 17 meters) over five decades in the late twelfth century, which reflects an estimated drop of between 10 percent and 12 percent in rainfall from the modern average. As I have described elsewhere, Tiwanaku was heavily dependent on raised-field agriculture nourished by streams and high groundwater levels. In a time of persistent drought, Tiwanaku's agriculture was no longer sustainable. The city collapsed and its population dispersed into much smaller communities and turned to the herding of native, drought-adapted llamas and alpacas and opportunistic dry farming.[6]

The same drought cycles affected a much wider area. Lago Argentino, in extreme southern Patagonia, lies in the southern hemisphere's westerly wind belt, which brings rainfall to the region.[7] Like the Sierra lakes, Lago Argentino rises and falls like a rainfall barometer. When the water rises, it buries the southern birch trees that have grown in the exposed basin. The birches grew for between a half century to a full century before perishing in about A.D. 1051. One hundred and twenty miles (200 kilometers) farther north, Lake Cardiel recovered from one of its lowest levels at about the same time, drowning shrubs rooted on the shore. Radiocarbon dates tell us they died between 1021 and 1228, remarkably close to the dates for the end of droughts in California.[8] As we saw in chapter 7, the mid-latitude storm track of the northern hemisphere remained north of California during these centuries, either because the circumpolar vortex (the circulation of the jet stream in the high troposphere) contracted, bringing warmer conditions north and south, or on account of a persistent ridge of high pressure. Both these conditions have occurred in modern times. Vortex contraction would also account for the droughts in southern Patagonia, but for different reasons. The Patagonian lakes lie to leeward of the Andes and in the zone where westerly winds are strongest. This would have had the effect of intensifying the Andes rain shadow (the area to leeward of a mountain range where much less rain falls) and triggering drought. The aberrant atmospheric circulation of medieval times brought much greater changes in rainfall than in temperature. This is certainly true of the Andes, where ice cores provide an unusually complete proxy of the rainfall shifts that affected not only Tiwanaku but also Chimor, the Chimu state of the arid north coast. However, unlike Tiwanaku, Chimor survived the same drought cycle and even thrived, which is surprising given the other climatic villain on the stage: El Niño.

El Niño, the so-called Christmas Child, has been part of the climatic dramatis personae in some of the earlier chapters, but it plays a major role in the Andean historical drama. Unlike droughts, El Niños are short-term climatic events, which bring torrential rainfall to the normally arid Peruvian coast. An El Niño can wipe out years of irrigation works in hours and reduce crop yields dramatically for months, if not years. El Niños also bring tropical sea currents that replace the cold, north-flowing Humboldt Current. Upwelling close inshore slows and the anchovies move away to cooler water. In the short space of a few months, the two foundations of coastal life crumble without warning, leaving thousands of people without sustenance.

In 1892, a Peruvian sea captain, Camilo Carrillo, published a paper in the *Bulletin* of the Lima Geographical Society, in which he described a tropical countercurrent known by local fisherfolk as El Niño ("the Child

El Niño, La Niña, and the Southern Oscillation

The Southern Oscillation is an irregular, seesawlike pattern of reversing surface air pressure between the eastern and western tropical Pacific. When surface pressure is high in the east, it is low in the western Pacific, and vice versa. Ocean warmings and pressure reversals are usually simultaneous. The British meteorologist Sir Gilbert Walker discovered the Southern Oscillation in the 1920s. He realized that when atmospheric pressure was high in the Pacific, it tended to be low in the Indian Ocean. The erratic swings of the Southern Oscillation changed rainfall patterns and wind directions in both areas. During the early 1960s, another meteorologist, Jacob Bjerknes, connected the Southern Oscillation to El Niño events, whence the term "El Niño/Southern Oscillation" (ENSO), which commonly appears in the scientific literature.

When warming occurs in the eastern Pacific, the sea surface temperature gradient between east and west declines. The trade-wind flow weakens, but the weaker east–west pressure gradient has to accompany the reduced trades. Such changes require pressure changes between the eastern and western tropical Pacific—the seesaw effect of the Southern Oscillation.

El Niño. Every few years, the northeast trade winds in the Pacific slacken and the inexorable forces of gravity kick in. Westerly winds increase over the Pacific east of New Guinea, generating subsurface Kelvin waves that push surface water to the east. The trade winds have piled up warm water in the west. As the winds slacken, that water flows eastward. More than 650 feet (200 meters) below the surface of the Pacific, the water temperature becomes much cooler. This thermocline is much closer to the surface in the eastern Pacific. When the Kelvin waves push east, the thermocline sinks in the east and warm water cascades above it toward the American shoreline. The Southern Oscillation changes direction and an El Niño is born. Now the cooler water is in the west. Searing droughts affect Australia and Indonesia, while rain clouds form over the arid Galapagos Islands and the Peruvian coast. The warm, moist air over South America causes the jet streams to lurch north, bringing storms to the Gulf of Mexico and heavy rain to California. The effects of a major El Niño can be severe on a global scale, as the map shows.

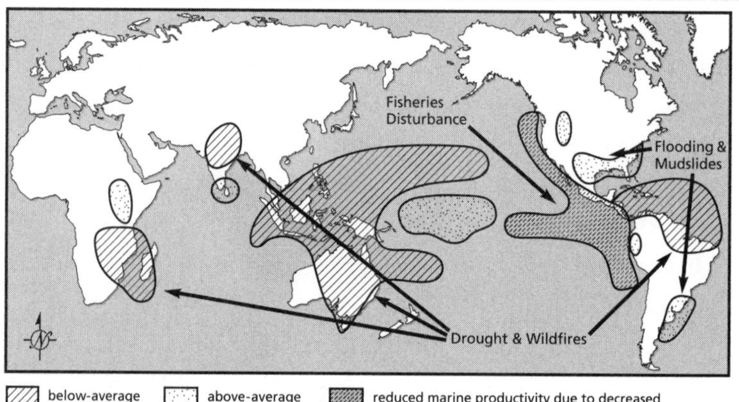

The global effects of a major El Niño; compiled from a variety of sources.

La Niña. Once warmer waters have spread into the central and eastern Pacific, some of the Kelvin waves rebound off the South American coast. The reflected waves eventually hit Asia and rebound again. Now the thermocline deepens in the west and shallows in the east. The easterly trades strengthen and the warm water pool in the western Pacific thickens. Upwelling renews, cooling surface waters in the east. El Niño gives way to cooler conditions, which in their extreme, and irregular, manifestations become La Niña, "the Young Girl," the cool and dry opposite of the Christmas Child. La Niña is still very much of a mystery, but it appears to last longer and have just as serious effects on human societies as its opposing twin, especially those subject to droughts caused by a cooler eastern Pacific.

The ENSO cycle is, and was, a powerful engine of climatic change, constantly swinging, totally unpredictable, and almost as powerful in its global impact as the seasons.

Jesus"), "because it has been observed to appear immediately after Christmas."[9] A flurry of scientific papers subsequently drew attention to what appeared to be a local phenomenon, which temporarily decimated the anchovy fisheries and brought heavy rainfall to the mountains and coast. El Niño remained a local meteorological curiosity until 1969, when the UCLA oceanographer Jacob Bjerknes linked atmospheric circulation in the Pacific with variations in sea temperatures in the tropical ocean. He linked El Niños with the seesawlike movements of the Southern Oscillation and identified them as not merely a local phenomenon but rather a global climatic force, now known to be second only to the seasons in its influence on global climate (for more about this, see the sidebar).

Despite increasingly sophisticated computer models, generations of climatologists have failed to establish any predictable pattern for ENSO events. Historical records over the 270-year period from 1690 to 1987 chronicle eighty-seven such events in Peru, spaced between two and ten years apart, but earlier than that the climatic record is still uncertain, partly because El Niños rarely last more than a couple of years and because they leave inconspicuous traces in the geological record, usually flood deposits in archaeological sites.

By all accounts, Chimor suffered from numerous El Niños. Chimu sites in the Moche and Jequetepeque valleys document massive floods. Dozens of Jequetepeque Valley excavations have documented four major floods, followed by extensive rebuilding, between 2150 B.C. and A.D. 1770.[10] A seabed core from a sheltered basin on the Pacific shelf 50 miles (80 kilometers) west of Lima shows a remarkably low concentration of El Niño events between A.D. 800 and 1250, although some did occur, evidenced in a lake in the Ecuadorian highlands.[11] From today's point of view, the significant one took place in A.D. 1230, when a great El Niño descended on the coast and caused widespread flooding and damage. These major events are but a few of the many ENSOs that affected coastal Peru over the centuries, each of them marked by thick flood deposits of rock and gravel carried by cascading water and preserved in river valley walls. There is also abundant evidence of dune activity during arid cycles, when fine sand blew over human settlements, preserving them in thick wind-blown layers in the same deposits. While droughts and El Niños occurred over wide areas, the effects were localized, with more damage in one place than in another, depending on local topography and other factors.

Droughts and El Niños were realities etched into every coastal farmer and fisher family's minds. Both arrived without warning and at irregular intervals. So did earthquakes and other tectonic activity that uplifted the coast suddenly and caused major landslides, some of which buried valley field systems. Strong winds contributed to desertification and caused sand dune shifts that smothered fertile land. The list of potential hazards was daunting. How, then, did Chimor adapt to such potential catastrophes?

AN EVENING IN A.D. 400. The sun casts long shadows across the plazas and pyramids of Cerro Blanco, near the modern city of Trujillo in the Moche Valley. A huge crowd of artisans, farmers, and petty officials fills the great courtyard in the shadow of the Pyramid of the Sun. A small thatched temple stands high above the plaza, its dark, open doorway in shadow. Drums beat; smoke from sacred fires swirls across the slope of the artificial mountain. Suddenly, the crowd falls silent, all eyes raised to the pyramid summit. The rays of the setting sun bathe a man clothed in gold and silver, who has emerged from the temple. He stands upright, scepter in hand, looking rigidly toward the western horizon. The polished metal glows fiery orange as it catches the slanting light; the living sun god has appeared before his subjects.[12]

We cannot understand the Chimu without going back a considerable distance in history, for many of their institutions went back centuries to the Moche, who ruled over much of the north coast during the early first millennium A.D.[13] The Moche kingdom covered the Lambayeque and several other valleys, probably as a series of fiefdoms governed by a small number of noble families with close kin ties. Political and economic power was in only a few hands, apparently those of men perceived to have supernatural powers. By the time the Chimu rose to prominence along the coast after 1100, the Moche were but a distant memory. The deeds of their mythic lords must have been richly embroidered in oral tradition.

The Moche state collapsed around A.D. 650, in part because of changing political conditions and competition from ambitious neighbors, but also because of the effects of a series of major drought cycles and massive El Niños. But their political institutions and religious beliefs seem to have survived, albeit in slightly altered forms. After A.D. 900, the center of political gravity shifted to another lordly dynasty, that of the Sicán. According to legend, the last Sicán lord, Fempellec, moved an ancient stone idol brought from afar by a royal ancestor, Naymlap, from the Lambayeque Valley to his capital at Batán Grande in the Leche Valley. Rains, floods, and disease promptly ensued. Fempellec's angry subjects cast him into the Pacific, and Chimu armies from the neighboring Moche Valley conquered the kingdom.

The Chimu had become a significant political power in the Moche Valley after A.D. 800. Their civilization had deep roots in the past, and among its sources were the by-then-almost-forgotten Moche. Perhaps the first Chimu rulers were descendants of once powerful Moche nobles, whose abandoned centers littered the coastal plain. Over the next four centuries, Chimor's lords began to wield expanding political authority. By 1200, they dominated a broad swath of the north and north-central Peruvian coast. Many of the cultural and political institutions forged by these people and their powerful lords, such as compulsory labor for the state and an elaborate road system, became part of the fabric of the Inca civilization that dominated the Andes when the Spanish arrived.

The Moche lords and their Sicán successors had invested in elaborate ceremonial centers, dominated by artificial mountains fashioned from adobe bricks. The last of these pyramid towns was Batán Grande, which fell before a huge El Niño as the Chimu conquered Sicán. Chimor's lords had apparently learned a lesson. They built an entirely different form of capital. Instead of investing in pyramids, they channeled resources into securing reliable food supplies. Rather than adobe mountains, they built themselves large, walled compounds where they dwelled in splendid isolation at their capital, Chan Chan, near the mouth of the Moche Valley.[14]

Chan Chan was an enormous city, one of the largest in the world in its heyday, rivaling London, Paris, and highland Mexico's Teotihuacán of centuries earlier. By 1200, Chan Chan's urban sprawl covered 7.7 miles (20 square kilometers), with only nobles, artisans, and other skilled workers living within the city core. No one knows the size of Chan Chan's population, but a staggering number of artisans lived in humble mud and cane huts along the southern and western edges of the central precincts—some 26,000 of them, including metalsmiths, weavers, and other specialists.[15] Another 3,000 lived close to the royal compounds; some 6,000 nobles and officials occupied detached adobe enclosures nearby. The city was apparently planning further expansion, for much of the land between Chan Chan and the Pacific remained open. How many people lived within the Chimu kingdom is unknown, but a figure in the 250,000 range would probably be a plausible estimate.

From their secluded compounds in the heart of the city, the lords of Chimor presided over an expanding but highly structured state. Egyptologists sometimes refer to the Nile Valley as an "organized oasis." The same term could be applied to the Chimor domains, centered as they were on coastal river valleys. We know little of the rulers or their system of governance, but Spanish accounts suggest that they governed by granting considerable, if carefully supervised, authority to local nobles. If the later Inca are any guide, then the lords of Chimor sent officials throughout their domains to inventory and supervise the activities of every household, and especially their fishing or agricultural production. The lords maintained their growing state by a combination of military force and tribute, helped by an efficient communication system. Everything flowed to the center, to major centers like Chan Chan, where artisans labored over fine gold and silver ornaments, feather headdresses, and other artifacts that denoted the prestige and power of their owners. Just like any other preindustrial civilization, the lords were careful to reward loyalty and prowess in battle with insignia and prestigious gifts. They were also well aware that their entire state rested on food supplies that could not be acquired by force or tribute. Without a stable agricultural basis, Chimor was very vulnerable indeed.

The Chimu capital was always at risk from El Niño floods and especially from lengthy droughts. Adding to the risk was a huge population of nonfarmers within the city. Apart from a small nobility and numerous artisans, numerous officials would have been required to administer the city, its markets, and its food supplies in a society where most authority was highly centralized. Fortunately, its rulers could draw on centuries of experience with irrigation, soil conservation, and water management. Their pragmatic strategies for survival were remarkably effective.

JEQUETEPEQUE VALLEY, A.D. 1200. Patchworks of green fields in a large, dry riverbed: the routine has varied little for generations. About every ten days, the farmers gather by the sluices that control water flow from the canal at the side of the valley. As a village official watches closely, the stone gates are opened. Precious water gushes from the canal and cascades through narrow channels into the fields. The meticulous irrigation work takes all day, with narrow defiles being opened and closed until the crops are watered evenly and fairly and each farmer has received the same allocation. Few words are exchanged as the hours pass, for everyone knows the routine well. They know the water is available, even if there has been no rain for months.

How did Chimor survive the jolting droughts between 1245 and 1310, at a time when El Niños arrived in unexpected waves?[16] Batán Grande's fate alone must have given Chimor's leaders ample grounds for worrying about food supplies for their growing city. For centuries, coastal farmers such as Moche villagers had used highly flexible agricultural systems, cultivating fertile plots on a small scale, laying them out along coastal hills where they could maximize runoff from springs and the occasional rainstorm. Such small-scale farming systems, which utilized as much land as possible consistent with available water supplies, worked well when population densities were considerably lower. The Moche strategy had the advantage of requiring relatively small-scale labor investment and no elaborate irrigation technologies, but there were now simply too many people, especially nonfarmers, for small-scale agriculture to support. In stark contrast, and confronted with rapidly growing cities and a burgeoning population, the lords of Chimor invested massively in closely organized, highly diversified agriculture. To do so, they relied on an ancient system of tribute by labor, known as *mi'ta*, which required annual labor by everyone to build major buildings and other public works.

Chan Chan depended on large, step-down wells for its domestic water supplies. The earliest public structures lay near to the ocean, where the water table was close to the surface. East of the city, low-lying terrain with a high water table allowed an elaborate complex of sunken gardens that extended 3 miles (5 kilometers) upstream from the Pacific. By 1100, Chimor's *mi'ta* laborers had also dug an enormous network of canals that watered the flatlands to the north and west of the city. Here irrigation water also replenished the urban aquifer. When a massive El Niño in that year devastated the irrigation system upstream of Chan Chan and altered the course of the Moche River, the ever persistent rulers commissioned the digging of a 43-mile (70-kilometer) intervalley canal to irrigate the land above the city with water from the neighboring Rio Chicama to the north. This expensive project was never completed, and the maintenance of water supplies required ever greater labor as the city expanded upstream into areas where wells had to be much deeper. The city eventually contracted toward the Pacific where the water table was shallower.

The Chimu created elaborate, redundant irrigation canals throughout their domains that supplied water to different parts of river valleys. Some of these canals were 18 to 25 miles (30 to 40 kilometers) long. The north side of the Jequetepeque Valley alone still features at least 250 miles (400 kilometers) of such canals built over many centuries. This vast canal system was never in use all at one time, for there was insufficient water to fill its entire length. The communities that depended upon it must have developed a carefully managed timetable of water delivery to everyone at different times. Today, local farmers water their crops about every ten days, and this was probably the practice in Chimu times as well. Those relying upon the network could bring different parts of the system online if some segments were destroyed by flooding or springs ran dry in drought cycles. The Chimu canal systems provided a practical way of mitigating extreme climatic uncertainty.

The Chimu also developed technologies that were designed to handle extremes of rainfall. The rulers of Farfán Sur, Cañoncillo, and other major centers in their domains constructed elaborate overflow weirs as part of their irrigation canals, especially for the aqueducts that bridged large canyons. When a major flood occurred, such overflows could slow turbulent water and prevent erosion. The same aqueducts boasted stone-lined conduits that allowed water to flow through the base of the structure without damaging it. Such measures were not totally successful, for there are many signs of rebuilding, but they certainly provided some protection against catastrophic surges.

The farmers also constructed crescent-shaped stone sand breaks in areas near the coast. These had the effect of slowing the flow of dune sand into irrigation canals and fields. Judging from abandoned breaks, this defensive strategy was apparently less effectual than their flood control measures.

In earlier centuries, the Moche had defended themselves against drought and flood by relying on extensive field systems that lay in different settings and could be maintained with relatively little labor. When El Niños brought floods, or during dry cycles, entire communities moved to different locations; field systems were rebuilt by each village. There was intense competition for fertile land from one valley to the next, with little attempt at centralized management of agriculture. The Chimu responded very differently, for they lived in a more densely populated world. Whereas their predecessors had lived in dispersed settlements across valley landscapes, they developed large towns and cities and practiced agriculture on a regional scale. Unlike the Moche, who developed flexible, less labor-intensive farming, the Chimu invested in entire agricultural landscapes created with enormous labor. For instance, Chimu lords built large storage reservoirs and terraced steep hills to control the flow of water down the slopes. Even in extreme droughts, their canals carried water from deep-cut riverbeds to terraces long distances away. With this infrastructure, the Chimu created thousands of acres of new fields; they brought water from great distances to harvest two or three crops a year where only one harvest had been possible before, and that at the time of the annual flood.

When reclamation became uneconomic, the lords of Chimor acquired land by conquest instead. As rulers often do, they rationalized this conquest, in this case developing an institution known to anthropologists as split inheritance. A lord would die and be buried in his compound. His mummified body would preside over his court as if he were still alive. Courtiers would attend him, talk to him as if he were alive, parade his body at public ceremonies. Meanwhile, his successor came to power with neither any possessions whatsoever nor a tax base in labor to support his establishment. His only option was to conquer new lands and their inhabitants. The constant military campaigns that resulted expanded Chimor from the Rio Santa to the Rio Jequetepeque, then to the Lambayeque and beyond, until the lords of Chan Chan ruled more than 700 miles (1,126 kilometers) of the coast. In agricultural terms, the strategy worked. At the height of their power, the Chimu controlled more than twelve river valleys with at least 125,000 cultivable acres (50,590 hectares), all farmed with hoes or digging sticks.

Chimu society was based on large cities, situated on alluvial fans, with adjacent hillside populations and an empire connected by a sophisticated road network. Chimu rulers were ruthless in their administration. It could not have been otherwise, given the vast investment in managing water supplies. They forced their subjects into large cities and severely restricted individual mobility. With such centralized control, the Chimu rulers could respond to environmental uncertainty and events like El Niños on a regional scale, diverting crops from one area to another, bringing unscathed irrigation canals online, and deploying large numbers of *mi'ta* laborers to repair aqueducts and canals.

Less than 10 percent of the coastal desert can be farmed, so Chimor also relied heavily on fishing. We learn from historical accounts that the fisherfolk of Chimor spoke their own dialects, married among themselves, and resided in separate communities under their own leaders. They were specialists, just like the farmers inland. They were largely self-sufficient, even growing reeds for their canoes in sunken gardens near the beach—also cotton for nets and lines, and gourds for floats. The fisherfolk traded their catches with farmers in exchange for produce. With a fishery at their doorstep that was accessible at least 280 days a year, Chimor's shore dwellers were, to a considerable extent, immune from drought. During El Niños they could catch fish sufficient for their own use, but upwelling slowed and often ceased during such events, so much so that anchovy catches were drastically reduced. The commerce in fish meal would have withered during such times. In a society of specialists, each community faced its own risks and developed strategies for doing so.

Unlike the Maya, whose civilization partially collapsed in the face of persistent arid cycles, Chimor, with its ruthless organization of valley landscapes, appears to have survived even prolonged droughts and exceptionally intense El Niños. With their growing cities and towns, the Chimu had no other option, faced as they were with the need to produce enormous food surpluses. Maya lords painted their commoners into an environmental corner, where a growing population, constant pressure to produce more food, and an overstressed rainforest environment were major factors in a political and social collapse. Chimor's cities survived the Medieval Warm Period because its lords closely supervised a subsistence culture that revolved around insuring against drought, floods, and deprivation. To this end, they created an elaborate organized oasis based on massive human labor and on draconian control. They thrived by cultivating different microenvironments, just as the Maya did. Both the Chimu and the Maya managed scarce water supplies with basically simple technology but a heavy investment in human labor. Both depended on a rigid social order and on rituals involving mediation between the living and supernatural worlds. In the case of the Maya, an inflexible ideology and the demands of the nobility, with their constant preoccupation with warfare, served to hasten the implosion of elaborate cities and an entire society in the face of persistent drought. The Chimu faced droughts that were, if anything, even more persistent than those in the Maya lowlands. However, there was a profound difference. Like the Maya, the lords of Chimor presided over a hierarchical and rigid society. But they lived in one of the driest environments on earth, where rainfall was rare and water came from afar. These realities preadapted farmer and ruler alike to lengthy dry spells as something experienced in every lifetime, to the point that they had no option but to diversify their food supplies and use every drop of water with sedulous care. And unlike the Maya, they had one of the richest coastal fisheries in the world at their doorstep, which enabled them to diversify their food supplies through elaborate irrigation works and the anchovy fishery. This they did with opportunistic success. The Chimu succeeded where the Maya failed simply because they lived with drought every day of their lives and had the hardwon experience of their remote ancestors to draw upon.

Chimor became a powerful force in the ever shifting political landscape of the Andes. Inevitably, formidable rivals rose in the highlands, who cast greedy eyes on the wealthy coastal state. The lords of Chimor controlled every aspect of their kingdom except one: the watersheds that provided mountain runoff to their valleys. In about 1470, ambitious Inca conquerors from the highlands gained control of these strategic water sources and conquered Chimor. The kingdom became part of Tawantinsuyu, "the Land of the Four Quarters," and its artisans were relocated en masse to the Inca capital at Cuzco.

The great droughts that ravaged human societies from the Great Basin to South America resulted from still only dimly understood interactions between atmosphere and ocean in the tropical Pacific. We must now journey there to explore what little is known about the warm centuries in the waters where ENSO affected the climate like a colossus.

CHAPTER 10

Bucking the Trades

I have heard from several sources, that the most sensitive balance was a man's testicles, and that at night or when the horizon was obscured, or inside the cabin this was the method used to find the focus of the swells off an island.
—Thomas Gladwin, *East Is a Big Bird* (1970)[1]

A.D. 1200. DAWN IN THE SOUTH PACIFIC. The crews of the large double-hulled canoes are weary. They have been sailing for seventeen days before a fitful westerly, the sails slatting back and forth in the lumpy swell. As darkness gives way to a gray dawn, the stars fade from the heavens. The weathered navigator scans the far horizon, sees nothing, and then gazes at the flocks of seabirds wheeling overhead. He stands still, legs apart, eyes closed, feeling through his feet the subtle gyrations of the waves bouncing off an invisible island over the horizon. Minutes pass; the motionless pondering continues. Then the pilot looks toward the horizon and points off the bow slightly to the right. His helmsman alters course, bringing the wind slightly off his right shoulder; the other canoe follows suit.

The sun climbs; the shadows on deck shorten. By noon, the wind has strengthened slightly. The canoes pick up speed, but the steersmen maintain the same angle to the swell. By now, the navigator has eaten some dried fish and drunk sparingly. He stands at the bow, looking into the far distance. Uneventful hours pass; the sun sinks to the west, casting the pilot's figure across the bow wave. Then, as sunset approaches, he silently raises his arm and points straight ahead. A distant cluster of trees stands out against the hard line of the horizon, just visible as the canoes rise and fall in the jumbled swell.

The canoes sail quietly through the night, aiming for the westerly end of the island, the crews ready to heave to if the land comes too close. At daybreak, huge palm trees stand tall on the low hills of the island. Everyone looks for signs of life, for village fires or signs of enemies. But the land is seemingly uninhabited. The pilot turns the canoes eastward, keeping a safe distance off, searching for a landing place. He spots a break in the cliffs, then a sandy beach. The sails are lowered, the canoes paddled to a safe berth in a place where high trees crowd on the landing place.

Unbeknownst to them, the crews of the two canoes have just completed one of the boldest voyages of exploration in history. They had taken a gamble, sailing eastward from Mangareva Island in eastern Polynesia to a landfall on Rapa Nui (Easter Island), in a direction normally closed to them by prevailing northeasterly winds. Some weeks before, the trades had dropped gradually. Calm, humid weather ensued. Then light winds filled in from the west and persisted far longer than the usual few days. So the navigator had sailed, not knowing what lay over the horizon. Days after the canoes landed on their new homeland, the northeast trades returned.

We think of Polynesia as a paradise world, a vast area of the central and southern Pacific encompassing a triangle whose points lie in Hawaii, New Zealand, and Rapa Nui and where life is little affected by climatic extremes and cycles of drought and flood. In fact, the South Pacific was at least as unpredictable climatically as other parts of the world, especially for canoe navigators and settlers on remote islands.

Rapa Nui is the remotest of all inhabited Pacific Islands. This makes it a special case in terms of both navigation and its subsequent history. Just to reach the island required unusual climatic conditions, for it lies

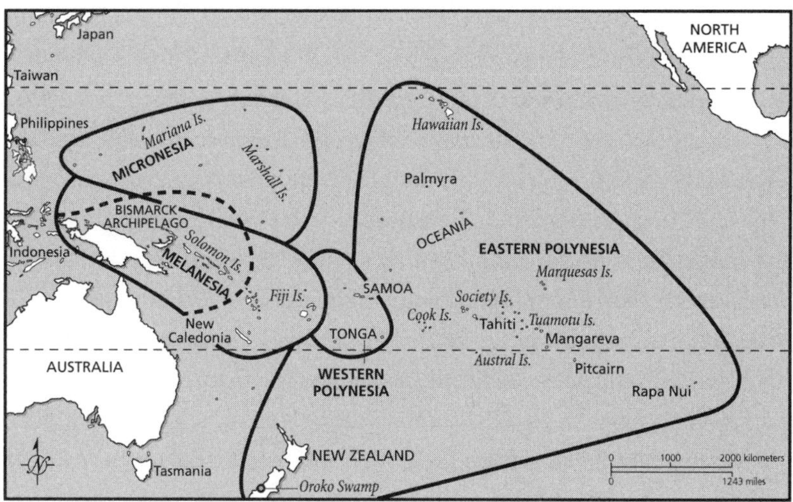

The Pacific, showing locations mentioned in this chapter.

in the teeth of the prevailing trade winds. Oral traditions from the island tell of a leader named Hotu Matu'a, the Great Parent, who colonized the island with his extended family. He arrived without such staples as pigs or dogs, which were an important element in Polynesian subsistence, although, of course, these might have perished after landing. Nor do we know whether there were later voyages to the island. The chances are that there were not, partly because long periods of westerly winds were highly unusual. As far as we can tell, too, there were no return trips to Mangareva or elsewhere, perhaps because the colonists rapidly deforested their new homeland, making it impossible to build new oceangoing canoes.

The settlers faced quite a challenge. Rapa Nui is subtropical, which means that the surrounding waters support no coral.[2] Fish are rarer than elsewhere in Polynesia; the rainfall is less and soaks quickly into the porous volcanic soils. Fresh water was hard to come by, but the islanders got by, cultivating sweet potatoes, taro, yams, and sugarcane. They brought chickens with them, which they raised in large stone chicken houses, but their diet was high in carbohydrates. Despite the challenges of an isolated environment, as many as fifteen thousand people may have lived on Rapa Nui in about 1600, relying on intensive agriculture and careful water management.

Like all Polynesian societies, Rapa Nui's had nobles and commoners. Oral traditions suggest that they divided the island into about a dozen kin-based territories, each with its stretch of coastline. The clans vied with one another in building platforms and colossal *moiae*, stone statues of high-ranking ancestors, which gazed out to sea. The archaeologist Jo Anne Van Tilburg has counted at least 887 *moiae* on the island, some of them still in the quarry. Over a period of four centuries, the islanders erected *moiae* and built platforms for them, a task that may have added about 25 percent to the food requirements of the islanders. The ecological cost was devastating. Between first settlement and about 1600, the descendants of the first settlers deforested the entire island of its magnificent indigenous palms. The island population declined by as much as 70 percent during the 1700s.

The settlement of Rapa Nui was the culmination of centuries of canoe voyaging through Polynesia. The ancient voyagers sailed from west to east across the Pacific in stages, from island to island, over many centuries. Many of their passages were no more than 300 miles (482 kilometers), but the voyage from Mangareva to Rapa Nui was about 1,600 miles (2,574 kilometers), a much more formidable proposition, with a tiny, and unknown, target at their destination. Just a few miles off course and they would have missed the island altogether and sailed on into deserted ocean. Remarkably, they also sailed eastward, against the prevailing trade winds, in canoes with limited abilities for sailing to windward. The Polynesians built their canoes with shell adzes and navigated confidently far out of sight of land by the stars, the winds, and ocean swells. To the Spanish explorer Ferdinand Magellan and his successors, the Pacific was a terrifying void that took months to cross, some 8,000 miles (12,874 kilometers) from the Strait of Magellan to New Guinea alone. The seemingly endless ocean was a "sea so vast the human mind can scarcely grasp it."[3] But to the Polynesians, the same ocean was a life-giving world teeming with fish, blessed with islands large and small where they could land, plant the soil, and raise their families. They lived close to its waters, were familiar with its every mood, and traversed its seaways with patience and consummate skill. But, like all of us, they lived, and sailed, at the mercy of powerful climatic forces.

Rapa Nui, Hawaii, and New Zealand were all colonized during the warm centuries, somewhat later than was thought until only a few years ago, which raises an interesting question. What effect did the climatic conditions of the Medieval Warm Period—a time of warmth and drought along much of the eastern Pacific Rim—have on the Polynesians? Was there, in fact, such a climatic event in the Pacific—and is there any significance in the apparent colonization of the most isolated landmasses of Polynesia during these centuries, especially since most long-distance canoe voyaging ended after 1500?

THE MEDIEVAL WARM Period is poorly documented throughout the Pacific; we cannot say with any confidence that it was a universal phenomenon. Only a handful of proxy records cover the past thousand years, most of them from tree rings or from growth rings in tropical coral. As of 2002, there were only three thousand-year-long temperature reconstructions from the southern hemisphere: one from Argentina; another from Chile; and a third from New Zealand. They are enough to tell us that temperatures and climatic conditions varied considerably from one area to the next. As always, climatic conditions were local, even if they were driven by much larger-scale atmospheric and global interactions.

The New Zealand sequence comes from Oroko Swamp on the west coast of South Island, where silver pines and other long-growing species flourish.[4] Silver pine wood is virtually impervious to decay, which means that numerous partially fossilized logs survive for tree-ring analysis. Edwin Cook and his research team developed a ring sequence that covers A.D. 700 to 1999, with a high degree of reliability back to A.D. 900. The record, meticulously checked for accuracy, documents two periods of generally above-average temperatures, in A.D. 1137–77 and again in 1210–60. There was a sharp and sustained cold period between A.D. 993

and 1091, the most extreme over the past eleven hundred years. There was rapid cooling after 1500, followed by modern warming.

The Oroko sequence documents temperatures that were between 0.6 and 0.9 degrees F (0.3 and 0.5 degrees C) warmer than the twentieth-century average for the region, and for 1210 to 1260, comparable to warming since 1950.

New Zealand enjoys a temperate climate, very different from that of Polynesia. The geographer Patrick Nunn has drawn on a broad range of climatic proxies, including an oxygen isotope analysis of a stalagmite from New Zealand, and sea level studies from various Pacific islands, to argue that the Medieval Warm Period was warm and dry, with persistent trade winds.[5] But he uses flow data from the Nile River on the other side of the world as an indicator of changes in the frequency of El Niño events to argue for a surge in their frequency around 1300. After 1250, Nunn argues, there was an abrupt change to cool, dry times, with increased storminess, that spanned the Little Ice Age, 1350 to 1850. Nunn calls this abrupt changeover the 1300 Event, a rapid event that precipitated a series of cultural and environmental catastrophes. According to Nunn, the 1300 Event saw significant disruptions in human settlement and subsistence, a decrease in voyaging, and greater competition, even endemic warfare throughout the Pacific.

Nunn's climatic evidence comes, for the most part, from observations of sea level changes and geological sediments that document erosion and flooding. Such climatic proxies are not nearly as accurate as tree rings or high-resolution coral growth data. Corals, in particular, have been the subject of intensive research in recent years and give us a very different picture of the warm centuries. They provide little support for the idea of a catastrophic climatic event throughout the region in 1300.

The paleoclimatologist Kim Cobb first came to Palmyra in 1998. To her delight, she located dozens of uneroded ancient coral heads on the atoll's west side. Since then, Cobb and her researchers have drilled cores from more than eighty such heads and revealed patterns of ocean temperatures and ENSO events over more than a thousand years. Her spliced-together cores now include samples from the Medieval Warm

Coral and Climate Change

The annual growth rings from recent coral samples from the central tropical Pacific extend back several centuries, and we know how to read these proxies so accurately that they rival the instrument records of modern times. They provide such invaluable data on past El Niños that climatologists have turned their attention to much older, long-dead fossil corals for data on the more remote past. Unfortunately, such corals are extremely rare, as most samples are fragments that have washed ashore in storms and escaped destruction by later gales. Given their tumultuous history after death, it's fortunate that any survive, but most of those that do are no more than a few decades long. They resemble tree-ring samples, and, like them, have to be pieced together into much longer master sequences based on dozens of short sections.

Why are corals such useful climatic indicators? They are formed by colonies of small marine invertebrates known as polyps, which secrete hard fortifications of calcium carbonate. Millions of polyps become fused together in elaborate structures, often with numerous branches. The polyps receive their nutrients from algae that live in these structures. Sunlight penetrates shallow water, shines on the algae, and creates their nutrients through photosynthesis. Corals cannot grow in dark, deeper water and require a narrow range of water temperatures between 77 and 84 degrees F (25 and 28.8 degrees C). They're fragile, too, usually surviving for no more than a few decades before storms rip them apart and fling them up on beaches, where they are pulverized. Intact, centuries-old coral is found only on islands in the less storm-plagued eastern Pacific with warm water, lush coral growth, and gales strong enough to detach coral heads and wash them ashore, but infrequent enough to leave them intact, undamaged by later bad weather. Few islands have yielded good old coral samples, except the remote and little visited Palmyra.*

The cores drilled into the Palmyra heads record growth rings in the corals, and through their isotopic content, variations in sea surface temperatures as the coral slowly grew. In cooler water, coral incorporates more of the heavier oxygen isotope O-18, whereas in warmer water there are higher counts of lighter O-16. By using high-precision mass spectrometry, paleoclimatologist Kim Cobb was able to measure small differences in the relative level of the two isotopes with remarkable accuracy. A difference of 0.02 percent in the relative level of the two isotopes signifies a temperature change of about 1.3 degrees F (about 0.6 degrees C).

Uranium-thorium dating techniques provided the ages of the dozens of small coral sequences from Palmyra that overlap one another. Uranium-thorium dating, sometimes called thorium-230 dating, is a radiometric dating technique that measures the age of carbonate materials such as coral. Uranium is soluble to some degree in all natural waters, so any coral precipitates minute traces of uranium. In contrast, thorium is insoluble and does not appear in the coral until the uranium-234 in it begins to decay to thorium-230. Uranium-thorium dating calculates the age of a coral sample by measuring the degree to which equilibrium has been restored between the radioactive isotope thorium-230 and its radioactive parent, uranium-234. Thanks to uranium-thorium dating, Cobb was able to piece together the beginnings of an important climatological archive.

The Palmyra cores show how El Niños exercise a powerful influence on Palmyra's climate, with warmer, wetter conditions during El Niños and cooler, drier conditions during La Niñas. During El

* Palmyra Island is a tiny Pacific atoll over 932 miles (1,500 kilometers) southwest of Hawaii. Fifty small islets with a total area of about 247 acres (100 hectares) form a horseshoe surrounding three lagoons. They stand but 6.5 feet (2 meters) or so above sea level, but the tall trees make the islands visible from 15 miles (24 kilometers) away on a clear day. A platform of coral and sand surrounds Palmyra, named after the American ship of that name, which sought refuge there on November 7, 1802. The Nature Conservancy purchased this near-pristine island in 2000.

Niños, O-18 levels in the coral fall, whereas they rise during La Niñas, providing a surprisingly reliable barometer of climatic events that can be dated accurately with uranium-thorium dates from dozens of short sequences that overlap one another. For instance, ENSO activity was reduced during the twelfth and fourteenth centuries compared with today.

Period, A.D. 928 to 961, 1149–1220, and 1317–1464, as well as for 1635–1703, the height of the Little Ice Age in Europe. A modern sequence extends from 1886 to 1998.[6] The variation in O-18 levels is relatively narrow, except for the tenth and mid-twentieth centuries. The tenth-century cold period was of some significance and appeared relatively abruptly, the coolest and perhaps the driest interval over the past eleven hundred years. The mid to late twentieth century was the warmest and wettest period of the past millennium. If the Palmyra corals are a reliable indicator, then a cool and dry La Niña condition prevailed over much of the eastern Pacific during a significant part of the Medieval Warm Period.

Palmyra's coral sequence is at present unique, except for a handful of more recent coral records. One such sample, from Australia's Great Barrier Reef, spanning 1565 to 1985, confirms the slightly warmer temperatures at Palmyra between 1635 and 1703, during Europe's Little Ice Age. The Great Barrier Reef sequence also shows that the tropical Pacific cooled just when the northern hemisphere warmed up during the late nineteenth century. Thus, there is good reason to believe that the Medieval Warm Period was a time of cool and relatively dry conditions over much of the Pacific.

Palmyra is but a tiny speck on a vast ocean, with relatively stable temperatures compared with the more marked shifts observed elsewhere in places like New Zealand. But its climatic history may have much broader significance. Cobb and her colleagues believe that changing sea surface temperature differences from east to west across the Pacific are a key player in global temperature patterns. Under this scenario, the gradient may have been greater during the tenth and twelfth centuries, bringing relatively cool and dry circumstances to this part of the Pacific during the Medieval Warm Period. Such conditions in the tropical central Pacific are very much part of the La Niña pattern, the seesaw opposite of El Niño, and the very same combination of factors brought severe droughts and lower lake levels to the American West and Mesoamerica, even to the Sahel in West Africa on the other side of the world. Furthermore, recent computer modeling relying on numerical experiments and well-established ocean-atmosphere models has replicated a period of more frequent ENSOs in the eastern Pacific during the Little Ice Age (which was warmer there), while the Medieval Warm Period saw reduced El Niño activity and cooler conditions over the same huge region.

Thousands of miles away, on the Pacific shelf off Lima, a core shows reduced ENSO activity during the period A.D. 800 to 1250, while in the Ecuadorian Andes Mountains, a high-altitude site at Laguna Pallcacocha has provided a twelve-thousand-year record of lake sediments from an area outside the Pacific, but one that is influenced by very much the same set of climatic conditions.[7] The cores from here provide evidence for El Niños in the form of marked debris layers caused by high rainfall, so the actual incidence may be much higher. El Niños can be detected as early as seven thousand years ago, but the pulse of events culminated between about 1200 and 1400, at just about the time of the final wave of East Polynesian voyaging.

The Palmyra research has overturned a long-held assumption that climate change in the tropical Pacific had followed that elsewhere, with a well-defined Medieval Warm Period followed by a cooler Little Ice Age. The opposite may have been the case in many areas. Almost certainly there were significant variations between different areas of the Pacific, perhaps measured by latitude. The so-called South Pacific Convergence Zone, a band of low-level wind, cloudiness, and rainfall that extends across the central Pacific from Vanuatu in the west to the Austral Islands in the east, was critical to human settlement. The zone moves and shifts in response to the ENSO seesaw. During El Niños, it moves northeast, while cyclonic activity shifts eastward and is more frequent. The zone shifts southwest during La Niñas, as well as responding to the Pacific Decadal Oscillation, a fifteen- to twenty-year fluctuation in sea surface temperature and rainfall whose warmer phases are associated with stronger and more frequent ENSO events.

With only a handful of meaningful climatic sequences, we still know little about the effects of the Medieval Warm Period on the vast reaches of the Pacific. Whereas in northern latitudes ice conditions or even a warmer shift of a few degrees could have significant effects, in the Pacific both climatic shifts and human settlement were affected more powerfully by the north–south movements of the Intertropical Convergence Zone, and especially by the gyrations of the Southern Oscillation and the El Niño and La Niña events associated with it. These shifts affected human societies in every region of the Pacific and far beyond. Everywhere in the Pacific, the global effects of the prolonged La Niña–like conditions that brought drought to many areas, higher rainfall to others, marked the warm centuries. And when La Niñas gave way to El Niños, as they did at times, and especially in the thirteenth century, the trade winds faltered and Polynesian navigators sailed eastward to new, remote lands.

On April 13, 1769, Captain James Cook anchored his ship, the *Endeavour*, in Matavi Bay. Cook's voyage from Plymouth in England to the heart of Polynesia took eight months, most of them out of sight of land. His landfall was a tribute to his remarkable navigational skills in an era when observing longitude was a novelty, but paled into insignificance alongside those of the Polynesian canoe pilots who had traversed the Pacific before him. Many centuries before the Spanish explorer Vasco Nuñez de Balboa gazed out over the Pacific from a "stout peak" on the Isthmus of Panama in 1513, island navigators had made their way across thousands of miles of open ocean to settle the remotest landmasses on earth.

Ever the seaman, Cook admired Tahitian canoes, used for trading, warfare, and long-distance voyaging. He wrote: "In these Proes or Pahee's as they call them from all accounts we can learn, these people sail

Indigenous Polynesian Navigation

When the British explorer Captain James Cook visited Tahiti in 1769, he puzzled over a question that still fascinates scholars: How had the Tahitians colonized their remote homeland? How had humans, with only simple canoes and none of the navigational tools used by Europeans, settled on the remotest islands of the Pacific? Cook met a Tahitian navigator, Tupaia, and asked him how canoe skippers made their way from island to island out of sight of land. Tupaia explained how they used the sun as a compass by day and the moon and stars by night.

Tupaia carried a mental file of Polynesia with him. He listed islands, the number of days required to sail to them, and their direction, which Cook made up into a rough sketch map. Modern scholars believe that Tupaia could define an area bounded by the Marquesas in the northeast, the Tuamotus to the east, the Australs to the south, and the Cook Islands to the southwest, an area of 1.6 million square miles (2.5 million square kilometers). Even Fiji and Samoa to the west lay within his consciousness.

No later explorers interviewed Tahitian navigators. Many scholars assumed the Pacific Islands had been colonized by canoes blown accidentally far offshore. But in 1965, the English cruising sailor David Lewis encountered aged canoe navigators in the Caroline Islands of Micronesia. He learned how, far from land, they used the zenith passages of key stars, as well as swell direction, waves reflected off distant land, and even the flights of sea and land birds to make landfall on island archipelagoes far from the departure point. These navigators were also able to return to their homes safely, using the same signs of sea and sky. Determined to preserve a rapidly vanishing art, Lewis sailed his European-designed, oceangoing catamaran from Rarotonga in the Cook Islands to New Zealand, using only a star map and a Polynesian navigator to help him. In the

1970s, Lewis apprenticed himself to the pilots of the Caroline Islands, learning how they made passages with the aid of sun, moon, stars, and cloud and swell formations, even passing birds.

In the late 1960s, the anthropologist Ben Finney began long-term experiments with replicas of ancient Polynesian canoes. Finney's first replica was *Nalehia*, a 40-foot (12-meter) copy of a Hawaiian royal canoe. Tests in Hawaii's windy waters showed that the vessel could sail across the wind, so Finney planned a voyage from Hawaii to Tahiti and back, using a replica built from a composite of known canoe designs from throughout the Pacific Islands. *Hokule'a*, designed by the Hawaiian Herb Kawainui Kane, is 62 feet (19 meters) long, with double hulls and two crab-claw-shaped sails. Finney, the Micronesian navigator Mau Piailug, and a mainly Hawaiian crew sailed *Hokule'a* from Hawaii to Tahiti and back in 1976. This journey was followed by a two-year voyage around the Pacific using only indigenous pilotage. Thanks to the successful *Hokule'a* experiments, ancient Polynesian navigational skills have been recorded in writing as well as oral tradition for posterity.

Every voyager took ample food supplies, usually dried root crops, sometimes even chickens and pigs. Dried fish were a staple, as were catches made while under way. Fresh water was carried in gourds and used sparingly, being replenished during heavy rain showers. Most canoes were provisioned to be as self-sufficient as possible, but the crews must have eaten and drunk sparingly to eke out their food and water for as long a time as possible. In reality, few passages lasted more than a couple of weeks, except when sailing far offshore to Hawaii and in eastern Polynesia.

in these seas from island to island for several hundred leagues, the Sun serving them for a compass by day and the Moon and Stars by night. When this comes to be prov'd we Shall no longer be at a loss to know how the Islands lying in these Seas came to be people'd."[8] He befriended Tupaia, an expert navigator, who carried with him a mental map of an area of the Pacific as large as Australia or the United States.

Five years later, on his third voyage, Cook lay at anchor off Easter Island, the most remote speck of land in Polynesia. By now, he had seen more of the Pacific and its indigenous inhabitants than any European explorer. Wherever he sailed, he witnessed remarkable similarities between island societies dispersed over thousands of miles. Cook marveled at their navigational abilities: "It is extraordinary that the same Nation should have spread themselves over all these isles in this vast Ocean from New Zealand to this island which is almost a fourth part of the circumference of the globe."[9] Having talked to Tupaia and other local pilots, he was in no doubt that the origins of the Polynesians lay in Southeast Asia and that their ancestors had sailed from island to island, from west to east. It is only in recent years that experiments with replica canoes and indigenous navigational methods have reconstructed ancient voyaging strategies.

HOW, THEN, DID Pacific settlement unfold and how did climate changes affect it? Thanks to an explosion of archaeological research over the past half century, we now know that colonization of the offshore Pacific began in the waters of the Bismarck Strait and the Solomon Islands in the far southwest, where a strong maritime tradition flourished as early as 1500 B.C. These "Lapita" people traded fine obsidian and other commodities over enormous distances, between islands hundreds of miles apart.[10] They were farmers, but above all sailors, who developed outrigger canoes with sails and used navigational techniques that allowed them to voyage far from land, something very different from the simple line-of-sight pilotage that had brought much earlier canoes to the

Solomon Islands from New Guinea and Southeast Asia before thirty thousand years ago.

Around 1200 B.C., long-distance voyages took Lapita canoes far beyond the Solomons, which had been the outer limit of human settlement for millennia. By 1100 to 900 B.C., Lapita families had settled in Fiji and in New Caledonia, also on Samoa. Within the brief compass of two to three centuries, a mere fifteen to twenty-five generations, Lapita seafarers had colonized islands over 2,800 miles (4,500 kilometers) of the offshore Pacific. Quite why they undertook these voyages of exploration and colonization, no one knows. Like the Norse, who suffered from land shortages and overcrowding, many Polynesians may have gone to sea because they had to. They lived in societies where property, house sites, ritual privileges, even esoteric knowledge were passed down to the eldest son. Junior siblings got the short end of the economic stick. Many of them struck out into the ocean in search of new lands to settle, where their families could find ample food.

Beyond Fiji and Tonga, the distances widen out to the Cooks and the Society Islands. There was, apparently, a pause before further voyages took people eastward from western Polynesia at about the time of Christ—the exact date is uncertain—and then on to the remotest islands of the central and southeastern Pacific during the late first millennium. Hawaii was settled by A.D. 800, New Zealand by 1000, perhaps earlier, and Rapa Nui by 1200, during the Medieval Warm Period.

IF THE LATEST radiocarbon dates are correct, the closing chapters of Polynesian voyaging unfolded during the centuries of the Medieval Warm Period. Was there, then, something unusual about passage-making conditions during the late first millennium?

Most of the time, the prevailing northeasterly and southeasterly trade winds blow on either side of the Intertropical Convergence Zone and are fueled by the Walker circulation, the powerful engine for El Niño events. As any small-boat sailor will tell you, sailing downwind in the trades is a paradisal experience. You sail at full speed for days on end, two jibs boomed out before the mast, a small part of the mainsail strapped in tight to dampen your rolling. Day and night, the ship sails herself effortlessly. You sit on watch at midnight in shorts and without shirts, the full moon high in the heavens. This is the ultimate in ocean sailing. But woe betide you if the winds reverse and blow from ahead. You slog miserably against lumpy seas and a nasty westerly, sick to your stomach and eager for the disturbance to pass.

Like modern yachts, Pacific canoes sail well with, and across, the wind. Modern replica canoes can sail at an angle of about 75 degrees to the wind (90 degrees is, of course, perpendicular to the wind).[11] Even sailing "full and bye," at a lesser angle to the wind, progress is slow. When pushed closer, the canoe slows down and slips away from the wind across its intended course. Like skippers everywhere before the days of internal combustion engines, Polynesian navigators waited for the trades to die down and for westerlies that would carry them in a direction that was normally upwind. Captain Cook, an expert seaman if ever there was one, well understood Polynesians' tactics for sailing to windward, for he himself was used to waiting for favorable winds. Tupaia told him that "during the Months of Novr Decembr & January Westerly Winds with rain prevail & as the inhabitants of the Islands know well how to make proper use of the winds, there will no difficulty arise in trading or sailing from Island to island even tho' they lie in an East & West direction."[12]

Today, we have mountains of data on the Pacific trades, which confirm what Tupaia told Captain Cook. There are periods during the southern summer when the trade winds falter and westerlies prevail. In fact, the winds blow from the southwest or southeast; this is ideal for east-sailing canoes, which perform best with the wind at an angle to the stern. The actual strategies for sailing east must have varied greatly, as many voyages would have begun in favorable westerly conditions, then experienced wind shifts as pressure systems moved across their course. In a remarkable series of experimental voyages, modern-day sailors have shown that eastbound voyages were entirely possible, provided that the sailors had patience and navigational expertise, and that they used indirect courses, always thinking ahead of their actual position.[13]

Westerly conditions are most prevalent during El Niño events. When the atmospheric pressure gradient over the Pacific reverses, the trades weaken and westerlies can blow for long periods, especially during the summer. These winds are usually strongest in the western and central Pacific, but they can reach as far as Rapa Nui and beyond. During major El Niños, canoes from Tonga could even have sailed direct as far as the Marquesas, where potsherds made of clay with minerals sourced to Fiji have been found.

For a long time, scientists assumed that the Medieval Warm Period brought higher temperatures and warmer conditions to the tropical Pacific, part of a global warming trend. Such conditions would have meant a higher incidence of El Niño events, commonly thought to be an accompaniment of warming. Recent studies of the Pacific trades have noted a weakening of the Walker circulation in response to warming. Did, in fact, the Medieval Warm Period bring more favorable voyaging conditions to the Pacific? This is a point of particular interest since the earliest known occupation of Rapa Nui, the remotest of all Polynesian islands, is now dated to A.D. 1200.

SUBTROPICAL RAPA NUI is the remotest inhabited landmass on earth, 2,300 miles (3,700 kilometers) from Chile to the east and 1,300 miles (2,100 kilometers) from Pitcairn Island to the west. The island covers 66 square miles (171 square kilometers) and is only 1,670 feet (510 meters) above sea level at its highest point, much lower than most Polynesian islands. It seems incredible that the Polynesians located this tiny dot in the Pacific, which is only some 8.5 miles (14 kilometers) across from north to south. The navigation is challenging enough with modern-day sextant and chronometer, even with a GPS. But the ancient canoe skippers located the island long before they actually sighted it, probably from the enormous flocks of nesting seabirds that live there, which could be sighted by an observant navigator as much as 186 miles (300 kilometers) before any land hove over the horizon.

The Dutch explorer Jacob Roggeveen sighted Rapa Nui on April 5, 1722, after a seventeen-day journey from Chile. He was astounded to be greeted by Polynesian speakers living on an island where the only watercraft were small, leak-prone canoes made of small fragments of wood stitched together with palm twine. Their homeland was treeless and poverty stricken, yet the inhabitants had managed to quarry and raise dozens of huge stone figures (*moiae*) that gazed silently out to sea. Where had the timber for seagoing canoes and for raising the statues come from? Half a century later, James Cook anchored off Rapa Nui. He described the islanders as "small, lean, timid, and miserable" and was equally puzzled by the lack of timber and of oceangoing canoes.

When did first settlement take place? The archaeologists Terry Hunt and Carl Lipo have recently excavated the only sand dune on the island, at Anakena. They dug through superbly preserved archaeological layers to a depth of nearly 138 inches (350 centimeters). There they hit sterile clay and a primordial soil that contained artifacts and the distinctive tubular root molds of the giant, extinct Rapa Nui palm. The same layer yielded the bones of numerous dolphins, which could only have been taken in deep water from oceangoing canoes. Eight radiocarbon dates from this level put the earliest settlement of the island at A.D. 1200, the height of the Medieval Warm Period.[14] The inhabitants were clearly able to fish from canoes far offshore and lived in a forested environment.

When the first canoes arrived, dense forests of huge Easter Island palms, some with trunk diameters of more than 6.5 feet (2 meters), covered the island. Within three centuries, the palms were extinct. So were twenty other tree species that once flourished on Rapa Nui. Two of the tallest trees, the toi (*Alphitonia zizyphoides*) and the evergreen (*Elaeocarpus rarotongensis*), grew to heights of 49 and 98 feet (30 and 15 meters) respectively, and were widely used in Polynesia for canoe hulls. All of these trees became extinct before 1500, as a result of intensive human exploitation of the thick forests for canoe timber, dwellings, levels and rollers for erecting *moiae*, and firewood, also from the depredations of imported rats. With no trees to build canoes, there was no way for the survivors of the resulting population crash to travel across the open Pacific to distant islands like Pitcairn and Mangareva, where deforestation also had disastrous consequences. The penalty of deforestation was complete isolation, for trade between Pitcairn and Mangareva ended by 1500, when commodities such as adze stone disappeared.

But how did canoes colonize Rapa Nui? The modern *Hokule'a* sailed from Mangareva to Rapa Nui in seventeen days in 1999, but she is a composite of several Polynesian canoe designs, with a relatively large sail area and an ability to work to windward that may have been superior to that of earlier outriggers. We do not know what earlier watercraft looked like, but they were certainly double-hulled and had rigs that were good across- and downwind. Their windward ability compared with the *Hokule'a*'s was probably limited, although we should remember that Norse *knarr*s turn out to sail much better to windward than people believed possible. Almost certainly, voyages of colonization were undertaken in periods of unusual westerly winds.

If ancient voyagers could sail in any direction, they would have left at any time. But with prevailing easterlies and canoes that could not sail well upwind, passages would have taken place sporadically. During El Niños, the easterly trade winds die down, and westerlies are common in January or February with an average rate of 3 to 5 knots (5.5 to 9.26 kilometers per hour). Such conditions would allow canoes traveling at a mean speed of 1.5 knots (2.7 kilometers per hour) to sail from western to eastern Polynesia in about twenty-two days at sea, if there were no calms or headwinds. Modern data suggest that the westerlies often did not extend far enough east during even major El Niños, but this does not mean they did not do so in the past. If they failed to discover land, they could always turn round and run before the trades, but we do not know whether the great voyages of discovery were one-way enterprises or made with the intention of returning. Much must have depended on political and social conditions at home, which might have made return impossible.

A double-hulled Tahitian canoe paddled by masked men, sketched by Philip de Bay, c. 1723. This is a small version of the much larger double-hulled canoes that sailed long distances. No pictures of such canoes survive, but this sketch gives a general impression of the type of hulls and sail.

The strongest El Niños occur during periods when they are most frequent, so the archaeologist Atholl Anderson and others believe that downwind voyaging to the east occurred during times of frequent westerlies. Polynesian navigators had no computer models to assist them, but they must have been familiar with the prolonged westerly conditions and high humidity that descended over the islands at irregular intervals. A voyage of any distance to the east and into the unknown would have required several weeks of westerlies, a far longer period than that required for a short interisland passage. As we have seen, there are signs that El Niños were more prevalent during the twelfth to fifteenth centuries, as the Medieval Warm Period ended and the first settlement of Rapa Nui took place.

Everyone agrees that ENSO events played a major role in ancient Pacific climate. The extent to which they affected Pacific voyaging is still intensely debated and we cannot yet be sure whether major El Niños really were significant factors in colonization. But we can be sure that the Polynesian islanders, like other people who base their lives around the sea, relied on an intimate knowledge of both maritime and terrestrial environments. If the Medieval Warm Period brought climatic conditions that were unusually favorable for long-distance navigation, these gifted sailors were ready to capitalize on them.

CHAPTER 11

The Flying Fish Ocean

The heat was intense, the thermometer indicated 108 degrees. A hot, blinding sandstorm filled our eyes and nostrils with microbe laden dust, and the all-pervading stench from putrefying bodies, impregnated clothes, hair and skin.

—Louis Klopsch, *The Christian Herald* (1900)[1]

"FAMINE IS INDIA'S SPECIALTY. EVERYWHERE famines are inconsequential incidents; in India they are devastating cataclysms," wrote a Victorian traveler who witnessed the horrors of the great Indian famine of 1896–99.[2] No one knows how many people died of starvation and famine-related diseases. A conservative figure is 1.9 million. The suffering beggared description. The journalist Julian Hawthorne was *Cosmopolitan Magazine*'s special correspondent in India. He arrived in the heart of the famine by train, shocked to see families of corpses seated under trees by the tracks: "There they squatted, all dead now, their flimsy garments fluttering around them, except when jackals had pulled the skeletons apart, in the hopeless search for marrow."[3] In Jubbulport (Jabalpur) in central India, American missionaries took him to the market, where he was horrified by the contrast between the plump merchants and the "bony remnants of human beings" begging for grain. He visited more famine victims in the poorhouse: "The joints of their knees stood out between the thighs and shinbones as in any other skeleton, so did their elbows; their fleshless jaws and skulls were supported on necks like those of plucked chickens. Their bodies—they had none, only the framework was left."[4] As the historian Mike Davis remarks, Queen Victoria's Jubilee in 1887 was "celebrated in carnage." The gross inadequacy of relief efforts by the British Raj contributed to the disaster.

The great Indian famines of the late nineteenth century were a direct result of a series of monsoon failures now known to be linked to major El Niño events. Such failures were nothing new. A thousand years ago, millions of people in South and Southeast Asia and along the shores of the Indian Ocean, from the Nile to China, lived at the mercy of the monsoon and its complex relationships with El Niño and La Niña.

As WE HAVE seen, there were prolonged La Niña conditions across much of the Pacific during the Medieval Warm Period, which had an effect as

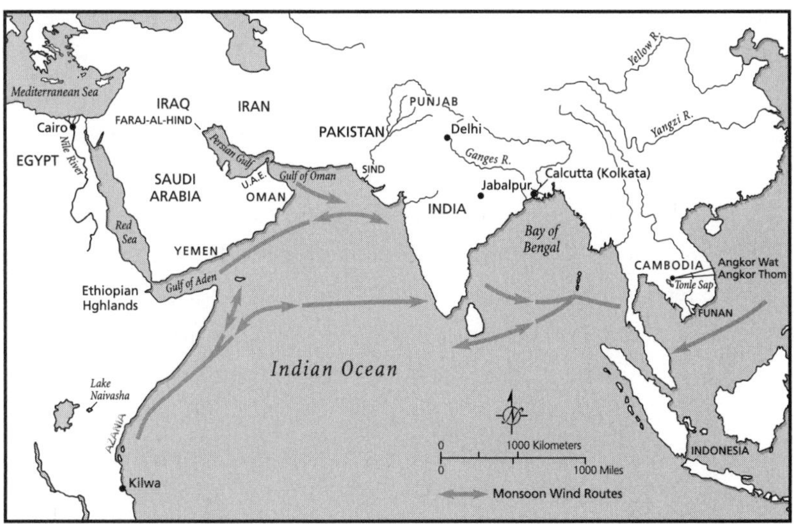

Locations mentioned in this chapter; also, general monsoon wind routes.

far away as northeast Africa, on the western shores of the Indian Ocean. Monsoon rains falling on the Ethiopian highlands provide about 90 percent of all the water flowing downriver during the river's annual rise. Contrary to popular belief, the summer Nile inundation is unpredictable and varies from year to year, as drought or plentiful rain affect Ethiopia. In the days before the Aswan Dam, Egyptian farmers went hungry when too little floodwater spilled across the floodplain. A flood that was 6 feet (2 meters) below average could leave up to three quarters of some provinces in upper Egypt without irrigation water. At the other extreme, an exceptionally high flood would rise precipitously and sweep everything, even entire villages, before it. With good reason, generations of ancient Egyptian pharaohs fretted about the flood levels and tried to predict them with carefully calibrated measuring devices, today called Nilometers.[5] So did their successors.

In A.D. 715, an Umayyad caliph, Sulayman Abd al-Malek, exercised about low river levels and the social disorder that ensued, ordered the construction of a Nilometer at the south end of Rhoda (Rawda) Island near Cairo.[6] A century and a half later, an Abbasid caliph, al-Mutawakkil, commissioned a major reconstruction under the direction of the Turkestan-born astronomer Abu'l 'Abbas Ahmad ibn Mohammad ibn Kathir al-Farghani, known in the west as Alfraganus. The great astronomer built an octagonal column within a stone-lined pit connected to the Nile by three tunnels. A scale of 19 Egyptian cubits (a cubit is 1.77 feet or 0.54 meters) carved on the column recorded the maximum and minimum levels of the river, making it capable of measuring a flood of 30 feet (9.2 meters). The walls of the Nilometer bear texts from the Quran referring to water, vegetation, and prosperity. An ideal inundation reached 16 cubits. Less meant drought and famine. A measurement over 19 cubits (33.6 feet [10.2 meters]) foretold catastrophic flooding.

The Rhoda column may seem like an unlikely source of climatological information, but its record extends back over fifteen hundred years. The archaeologist and demographer Fekri Hassan has calibrated flood data from Rhoda to reflect silting and other factors. He records a major episode of low flood levels between A.D. 930 and 1070, followed by a high from 1070 to 1180, after which the Nile plunged again, with low inundations for 170 years. With poor floods came meager harvests, famine, and inflated grain prices. The caliphs had ample cause for concern and good reason to invest in their Nilometer. Between A.D. 622 and 999, there were 102 years with poor floods—that is, the floods were poor in just under 28 percent of years.[7] In A.D. 967, 600,000 people died of starvation and famine-related diseases, a quarter of Egypt's population. During another famine, in A.D. 1220–21, between 100 and 500 people a day perished in Cairo alone.

The prolonged tenth- and thirteenth-century droughts known from the Rhoda Nilometer also affected East Africa. Lake Naivasha in central Kenya experienced a long period of intense aridity from about A.D. 1000 to 1270, with only one brief period of higher rainfall, from about 1200 to 1240. In contrast, the period from 1770 to 1850, the apogee of the Little Ice Age, was generally wetter.[8] The Naivasha record is far from unique. Lakes Victoria, Tanganyika, and Malawi all experienced prolonged droughts and low water levels after 1040. Mount Kilimanjaro experienced unusual aridity during the same century. The lakes lie outside the East African highlands, where cattle people had flourished for at least two thousand years. We know nothing of these groups, who had few possessions and were constantly on the move, but when lengthy droughts arrived, the people would have stayed within easy reach of permanent water supplies. Judging from historic droughts, they probably lost thousands of head of cattle as grazing grass withered year after year. However, like other pastoralists, they would increase the size of their herds during well-watered years to cushion themselves against future potential losses. As with the herders of the Saharan Sahel, the fate of their herds depended on climatic forces generated on the other side of the world.

About a thousand years ago, just as the African droughts were at their height, Muslim traders established themselves in small communities along what is now the Kenyan and Tanzanian coast. They had come in search of ivory, timber, and tropical products for many centuries, but now they settled permanently on the East African shore. The attraction must have been purely commercial, for the coastal environment is hot and dry. An arid and inhospitable hinterland separated these towns from the herders of the far interior, but the small, polyglot "stone towns" (so named after their coral houses), partly African, partly Islamic, prospered off a trade in African gold, iron, ivory, and timber. They were the outposts of the vast commercial world of a thousand years ago that depended on the monsoon and on a classic sailing vessel, the lateen-sailed dhow.[9]

THE NORTHERN INDIAN Ocean is, on the whole, the kindest of the world's great seas to the mariner. Ancient Arabic songs called it the Flying Fish Ocean. Much smaller than the Pacific and with more reliable winds, the Indian is an embayed ocean, checked by Asia and divided in the north by India into the Arabian Gulf and the Bay of Bengal, the latter linking with the eastern seas off Southeast Asia. Asia confines the ocean and upsets the normal passage of ocean winds. North of the equator, the mariner sails in the monsoon belt. On the western side of the Indian Ocean, from the Mozambique Channel between Africa and Madagascar, north and east through the Gulf of Arabia and into the Bay of Bengal and adjacent waters, the rhythms of the northeast and southwest monsoons have governed ocean voyaging for thousands of years.

During the first century A.D., an anonymous Greek-Alexandrine merchant or sailor compiled the *Periplus Maris Erythraei*, "The Periplus of the Erythraean Sea." The Greek word *periplus* means "a going about," which is just what the unknown writer recorded, almost certainly from firsthand knowledge of an ocean that many skippers sailed in all weathers and at all times of year. He had sailed from port to port from Africa to India, and also direct far offshore, on the wings of the monsoon winds: "This whole voyage as above described, from Cana and Eudaemon Arabia, they used to make in small vessels, sailing close around the shores of the gulfs; and Hippalus was the pilot who by observing the location of the ports and the conditions of the sea, first discovered how to lay his course straight across the ocean."[10] The Book of Revelation de-

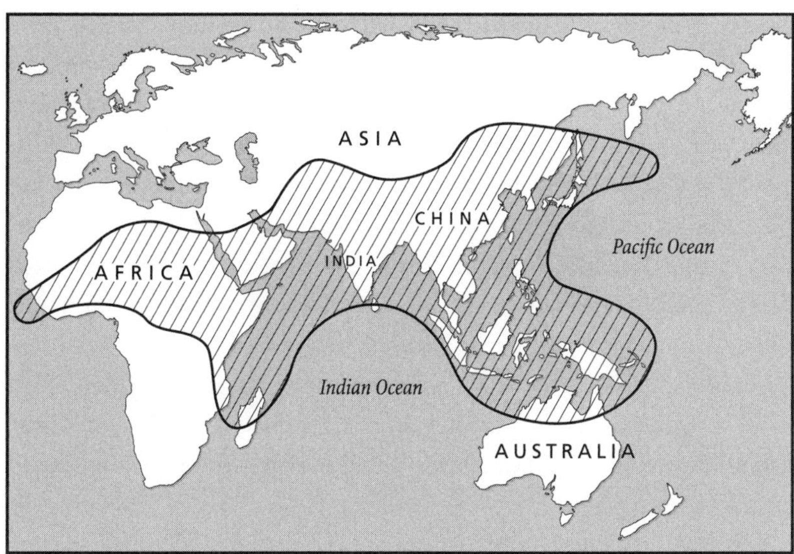

The extent of monsoons.

scribes a Red Sea and monsoon trade in lyrical terms: "Gold, and silver, and precious stones, and pearls, and fine linen, and purple, and silk, and all sweet wood, and all manner of vessels of ivory."[11] This was the trade that led to the founding of strategic commercial towns on the East African coast as far south as Kilwa in southern Tanzania.

Each monsoon wind blows for about half the year, but the changeover is never abrupt. The southwest monsoon brings rain to the west coast of India by the end of May, reaches its full and boisterous strength during July, and lessens until October, when it disperses. Heavy rainfall and strong winds effectively close exposed ports on the Indian coast even for large sailing vessels during strong monsoon years, often associated with La Niña–like conditions in the Pacific. The rains can fall for forty or fifty days with only short periods of fair weather. These are not gracious conditions for the ocean sailor, especially not for coasting vessels with open cargo holds.

The northeast monsoon is benign and dependable, never reaches great strength, and blows day and night with a predictable routine that is balm to a skipper's soul. Between November and May, this remarkable breeze carried dhows from the Persian Gulf to India, and vessels from Indian ports to Mesopotamia and far westward along the shores of southern Arabia to the distant, spice-rich island of Socotra at the mouth of the second great arm of the Indian Ocean, the Red Sea. Meanwhile, as merchant ships reached India from the east, Chinese junks and other ships bound for the Malay Peninsula and points east had already left port on the southwest monsoon that blew across the Bay of Bengal from April to July. The same winds could carry sailing vessels as far as the coast of Vietnam and into the South China Sea until September, when the weakening southerlies gave way to the northeasterly monsoon. This, in turn, lasted until April, when the cycle renewed itself. The monsoon winds allowed India and China to establish contact with the Roman and Islamic worlds by sea.

Just like those of the Pacific Ocean trades, the patterns of the monsoon winds were far from constant. An irregular and still little-understood cycle shifted monsoon winds north and south over the Indian Ocean. When the winds were to the south, the Ethiopian highlands received ample rainfall, but when they moved northward, rainfall diminished, culminating in severe droughts lasting a decade or more. The droughts of the warm centuries were especially prolonged, apparently during still little-documented periods of La Niña conditions in the Pacific. During the tenth- and thirteenth-century arid cycles, the monsoon winds were to the north. As we have seen, Ethiopia and the East African lakes region experienced diminished rainfall. So, while India received heavy rain and the southwest monsoon blew strongly, Africa suffered under drought. According to the historian Ian Blanchard, the north–south cycle had a periodicity of about 100 to 120 years, operating like a slow-moving pendulum.[12]

Sailing conditions in the Indian Ocean were challenging during the warm centuries, when strong southwesterly monsoons, associated with persistent La Niña conditions farther east in the Pacific and blowing onshore, made the approach to India's western coast a dangerous proposition. The polyglot skippers who plied the Red Sea, Persian Gulf, and Indian Ocean adjusted their sea routes as the prevailing winds changed.

So did the merchants, who could save days by traveling overland from ports of the Red Sea to the Nile instead of battling strong headwinds. Travel by land or sea was never easy. "By day . . . many [ships] are lost because the straits are stormy because of land winds," wrote the Portuguese traveler Thomé Pires in 1513.[13] The strong southwesterly monsoons of the Medieval Warm Period made for fast ocean passages eastward, which may have made the East African coast, with its rich stocks of elephant ivory, especially attractive to long-distance sailors. Permanent settlement by merchants with connections elsewhere may have resulted. (The ivory of the African elephant is softer and more easily carved than that of its Indian relative, a fact that created an insatiable demand for African ivory in India, where it was prized for, among other things, bridal ornaments.)

Sailing conditions during much of the year may have been more hazardous, but, from the dhow skipper's perspective, the situation in the Indian Ocean during the warm centuries was somewhat like that faced by camel caravans in the Sahara. When rainfall increased slightly in the desert, the caravans used routes in the heart of the Sahara where water could now be found. During dry times, they moved westward, closer to the Atlantic. Camels allowed their owners to adapt effortlessly to climatic shifts. So did the dhow, which sailed well downwind in the hands of sailors who waited patiently for wind shifts and listened carefully to intelligence about the north–south movement of the best sailing routes. Their ability to adapt quickly to different wind conditions meant that the Indian Ocean trade never ceased through the warm centuries. In the days before automobiles, railroads, and steamships, the tentacles of long-distance trade on land and ocean were infinitely adaptable and constantly shifting. On the shores of the Indian Ocean, the ports where ships called might silt up; pirates might infest a once safe estuary; the pattern of the monsoon winds might falter. Visiting ships would now trade elsewhere.

LOW NILE FLOODS, drought in East Africa, strong southwesterly monsoon winds, and changing sailing routes across the Indian Ocean: all these phenomena stemmed ultimately from climatic shifts across an enormous sweep of the globe and from an area of the southwestern Pacific known as the Hot Pool.

We've examined many of the rainfall and temperature changes in earlier chapters. Far from the Indian Ocean, a deep-sea core in California's Santa Barbara Channel chronicles a long period of cooler sea surface temperatures and strong upwelling from about A.D. 500 to 1300 that is typical of La Niña, the counterpoint of El Niño (as described in chapter 7). The ancient corals from Palmyra Island in the central Pacific also hint at relatively cool and dry, perhaps sometimes La Niña conditions, during the twelfth century (see chapter 10). A deep-sea core off coastal Peru documents a long period of La Niña–like climate between A.D. 800 and 1250. Lakes in the Ecuadorian Andes and the central Chilean lowlands also record a low incidence of the floods and torrential rainfall typical of strong El Niños between A.D. 900 and 1200 (as we saw in chapter 9).

The trajectory of tenth-to-thirteenth-century drought carried across a then cool and dry Pacific into the area of the Hot Pool, which is the crucible for the monsoon climates of Asia and the Indian Ocean. This reservoir of warm water sloshes over an area east to west along the equator for some 9,000 miles (14,500 kilometers) and about 1,500 miles (2,400 kilometers) north to south—an enormous bathtub, as it were, that covers an area four times that of the United States. The Hot Pool includes the waters of the western equatorial Pacific between New Guinea and Samoa, with a long tail extending through the Indonesian Archipelago and far into the Indian Ocean. These are the warmest waters on earth, warm enough to drive heat and moisture high into the atmosphere and to affect the climates of neighboring landmasses—China and India. The slow and cyclical fluctuations in the size and temperature of the Hot Pool may be closely linked to the intensity of El Niños, but no one yet knows what causes them.

When El Niños form, the Hot Pool moves eastward, closer to the international date line. A strong high-pressure system builds over Indonesia. The high-pressure center delays the monsoon, drought ensues, and forest fires rage over wide areas, as they did memorably during the great El Niño of 1997–98. Strong ENSOs reduce the amount of rainfall over much of Southeast Asia and as far south as Australia and New Zealand. During the intense El Niño cycles of 1870–1900, New South Wales and Victoria in Australia turned into dustbowls, with huge forest fires and dust storms lasting for days. Millions of sheep perished; crops failed over a wide area. But every El Niño is different, so the effects vary each time, modified also by still little-known decadal and longer shifts in such phenomena as Eurasian snow cover.

The climate of the Indian Ocean interacts with the Pacific's ever shifting Southern Oscillation, but other, quite independent variables make a significant difference. One certainly cannot claim that there is a direct and invariable connection between a strong El Niño and monsoon failure over South and Southeast Asia. But the links between ENSOs and Indian droughts are real. Twenty out of twenty-two El Niño years between 1870 and 1991 saw drought or lower than average rainfall. There's no reason to believe that such linkages weren't operating in the past.

What, then, was the result of the persistent cooler conditions that flourished over the Pacific for much of the tenth to thirteenth centuries, not continuously, but certainly as the dominant climatic pattern? During cooler periods and La Niñas, the Hot Pool moves westward, away from the international date line. Unimpeded monsoons spread across Southeast and East Asia, bringing ample rainfall, sometimes too much.

We still know little about the workings of the Hot Pool, but it's clear that even minor changes in sea surface temperatures can have a significant effect on climate in surrounding areas. In general terms, there seems little doubt that cool, arid La Niña–like conditions mean stronger summer monsoons and higher rainfall in South and Southeast Asia, even if the correlation is not exact or invariable. A few climatic proxies scattered over a vast area hint at a wetter period with stronger summer monsoons from A.D. 1000 to 1350. Among them are stalagmites in a shallow cave in Oman, a deep-sea core off Pakistan, and a fossil pollen sequence from Maili in northeast China.

These three and a half centuries were times of considerable turmoil in India, marked by Islamic incursions from northern nomads, the founding of a Muslim dynasty in Delhi, and the expulsion of Buddhism. But apparently, the generally prevalent La Niña–like conditions in the Pacific ensured good monsoon rains much of the time.

IN INDIA, THE monsoon is much more than a matter of meteorology. Throughout the subcontinent, human existence, the very fabric of daily life, unfolds around two seasons, the wet and the dry. The wet season brings warm, moist conditions and heavy rain, carried by the monsoon winds blowing inland from the ocean. The other half of the year, the arid season, enjoys cool, dry air from the north. The coming of the summer monsoon is a highlight of the year to those who have suffered through the buildup after the pleasant winter months—weeks of torrid heat. Wrote Colonel Edward Tennant of the East India Company in 1886: "The sky, instead of its brilliant blue, assumes the sullen tint of lead. . . . The days become overcast and hot, banks of cloud rise over the ocean to the west. . . . At last the sudden lightnings flash among the hills, and shoot through the clouds that overhang the sea, and with a crash of thunder the monsoon bursts over the hungry land."[14] The day when the monsoon broke was always memorable. The sixth-century writer Subandhu wrote: "Peacocks danced at eventide."

The Indian summer monsoon arrives in June as the land absorbs heat from the sun faster than the ocean does. Air masses over the land heat up, expand, and rise. As the air rises, cooler, moister, and heavier air from the ocean replaces it. The winds shift to the southwest, blowing inshore, bringing heavy rainfall. Central and western India and Pakistan receive more than 90 percent of their rainfall during the three months of the summer monsoon. The south and northwest receive 50 percent to 75 percent of their annual totals from the monsoon. In the semiarid regions of South Asia, the fluctuations of monsoon rainfall can make the difference between life and death. Very often, however, it's not a lack of rainfall that is the problem, but the rain's timing. A monsoon season may start with a deluge; then no more rain may fall for the rest of the year. India is a high-risk environment for farmers, as is northern China (described in chapter 12).

A thousand years ago, India was a patchwork of forest and irregularly cleared lands. Political events or war could result in the complete abandonment of large areas of once cultivated land, which became open reaches of grassland. Thousands of years of human interference had devastated the natural vegetation. For example, had farmers not cleared brush and set fires, much of the western coastal region would have been covered with tropical evergreen forest or dry tropical deciduous or thorn woodland. In 1837, the British surveyor W. H. Sykes wrote of the countryside around Mumbai (Bombay) that during the dry months of April and May "the country appears an arid desert. After the monsoon however . . . the country appears one great field of grain."[15] To survive as a farmer in an environment of such contrasts, where intense heat ravaged the land for months on end and most rain fell within the compass of a few months, required patience and adaptability, also mobility.

As so often happens, we know nothing of the fortunes of the millions of farmers who labored, throughout the warm centuries, in the shadow of nomad raids and indiscriminate plundering. Their settlements have long vanished, many buried under deep accumulations of river silt. As happened everywhere, droughts and floods were ignored in the chronicles of the day. This is hardly surprising, for the warm centuries were a time of considerable upheaval. While Europe was entering the High Middle Ages and the Maya wrestled with drought, India was encountering Islam. The commander of the first Muslim army to reach semiarid Sind, in what is now Pakistan, was not impressed. He reported that "water is scarce, the fruits are poor, and the robbers are bold; if a few troops are sent they will be slain, if many, they will starve."[16] For centuries, raids in search of plunder, especially from the Ghazni kingdom in Afghanistan, ravaged India. The plundered wealth from these incursions turned Ghazni into a great center of Islamic learning during the eleventh century. Inevitably, raiding led to conquest in the end. After 1206, Muslim dynasties ruled from Delhi for 320 years. The sultan

Shams-ud-din, who ruled from 1211 to 1236, was able to keep Ginghis Khan at bay through consummate diplomacy.

These events unfolded at a time of prolonged cool and La Niña–like conditions in the southwestern Pacific, which would have led to good monsoon rainfall most years. No farmer could relax, however, for droughts could descend for several years without warning, as a result of ENSO shifts in the southwestern Pacific, before cooler conditions there brought plentiful monsoon rains once again. For this reason, there must have been occasional serious, and long forgotten, periods of drought-caused hunger. We have only the testimony of later centuries to inform us.

Even after the warm centuries, crop failures plagued India. Before railroads and improved communications made prompt grain shipments possible, famine was endemic. In 1344–45, such a severe famine affected India that even royalty starved. Writing of northern India, the sixteenth-century Mughal emperor Babur wrote that villages and even towns were "depopulated and set up in a moment! If the people of a large town, one inhabited for years even, flee from it, they do it in a way that not a sign or trace of them remains in a day or a day and a half. On the other hand, if they fix their eyes upon a place in which to settle, they need not dig water-courses, or construct dams because their crops are all rain-grown."[17] The failure of the monsoon rains in 1629 and again in 1630 depopulated entire rural districts. Millions of people and their cattle perished. Cholera epidemics carried away entire villages. Another major drought came in 1685–88. A century later, the great hunger of 1770 depopulated and ravaged a third of Bengal. The South Asian monsoon failed again in 1789, and the failure was followed by intense droughts that descended on Australia, Mexico, and southern Africa in 1790. Six hundred thousand people starved to death in the northern Madras region in 1792. The dead and dying blocked the streets of Calcutta. One reason for the high casualty rates: until the eighteenth century, most of India relied on high-risk "dry" agriculture. Irrigation works did not become commonplace except near permanent rivers until the eighteenth century.

The monsoon also played an important role in the flamboyant Khmer civilization of Southeast Asia. Funan was the medieval Chinese name for the lower Mekong River. Its delta was a land of wealthy kingdoms and settlements surrounded by high earthworks, where the Chinese obtained bronze, gold, and spices.[18] By the beginning of the warm centuries, the center of political and economic power had moved upstream to the Tonle Sap, the central basin of Cambodia. Tonle Sap ("Large Freshwater Lake") is a large shallow lake during the dry season covering about 1,150 square miles (3,000 square kilometers) and some 40 miles (66 kilometers) long. A river of the same name connects the lake with the Mekong. During the monsoon rains between August and October, so much floodwater pours into the Mekong that the river reverses its flow and backs up into the Tonle Sap. The lake swells rapidly, flooding surrounding fields and forests, and eventually covering as much as 6,200 square miles (16,000 square kilometers) with depths up to 30 feet (9 meters). The Tonle Sap is now between 80 and 120 miles (133 to 167 kilometers) long and up to 30 miles (50 kilometers) wide. The fish from the Tonle Sap breed in the flooded forests along the edge of the lake. Many of them swim out through the outflow into the Mekong. Late in October, the floodwaters slowly retreat, trapping millions of fish in muddy bayous. The Tonle Sap's bountiful environment was a paradise for rice farmers and generated enough food to support a glittering, wealthy civilization—provided the local rulers built water control systems and thereafter managed water supplies.

Tonle Sap was a cockpit of competing lords and internecine warfare for many centuries. But in A.D. 802 a dynamic Khmer monarch, Jayavarman II, defeated his competitors and carved out the Angkor state, held together by Hindu beliefs, plus force and tribute payments. Jayavarman consolidated his kingdom by proclaiming himself a god-king; his subjects worshipped him as a deity. All the resources of an increasingly centralized government were devoted to the cult of the divine monarch. Everyone, whether general, noble, priest, or commoner, was expected to subordinate his or her ambitions to the need to perpetuate the existence of the king on earth and his identity with the god in this life and next. Jayavarman II ruled for forty-five years, the first of at least three dynasties of Khmer kings who presided over a state that reached the height of its prosperity during the ample monsoons between A.D. 900 and 1200, the Medieval Warm Period.

Jayavarman II and his successors presented themselves as reincarnations of the Hindu creator god Siva: the *varman*, or protector. Under them, a tightly controlled bureaucracy of high-status families supervised every aspect of Khmer life and owned land farmed by others. Local agriculture produced large enough food surpluses to support an immense and long process of temple building. During the dry months, the entire kingdom devoted itself to building ever more splendid palaces and temples constructed on artificial mounds in the hub of the Khmer universe, an area known today as Angkor.

The temples of the Khmer rulers dwarf those of Egyptian pharaohs or Maya lords. Four years after his accession in 1113, King Suryavarman II commenced building Angkor Wat, a masterpiece of beauty, wonder, and magnificence. Every detail of this extraordinary structure reproduces part of the heavenly world in a terrestrial mode—a central continent known as Jambudvipa, with Meru, the cosmic mountain, rising from its center, the highest tower at Angkor Wat. Four lesser towers represent Meru's lesser peaks, the enclosure wall depicting the mountain at the edge of the world; the exterior moat, the ocean beyond. Yards of superb bas-reliefs depict Suryavarman receiving officials, progressing through a forest on an elephant accompanied by heavily armed soldiers. Celestial maidens, slender and sensuous, dance with the promise of the delights of paradise.

A later monarch, Jayavarman VII, erected a huge new capital at nearby Angkor Thom in 1181. He and his immediate successors continued to spend without constraint. He dedicated the Ta Proehm temple to his mother. An inscription recorded that about twelve thousand people worked for the temple, living off rice grown by sixty-six thousand farmers, an indication of the scale of effort that went into maintaining what can only be described as a weirdly centripetal kingdom.

In every preindustrial civilization, like, for instance, those of the Egyptians and the Maya, everything flowed to the center, for the rulers controlled the labor of the ruled. The Khmer empire was an extreme example of such centralization, for everybody and everything was devoted to the divine king and his celestial immortality. Khmer rulers collected taxes in labor and grain, imposed tribute, and annexed the labor of their subjects to build their stupendous temples, whatever the cost. As a result, their centripetal domains lived on the knife-edge of sustainability.

The kingdom balanced on a carefully engineered water system fed by the annual monsoon rains.[19] The floodwaters helped rice production along the edge of the lake, but the water management network controlled water coming from the hills, as well as from the heavens. Angkor Wat and other Khmer temples mesmerize archaeologist and tourist alike, but it is only in recent years that scientists have gained a comprehensive and detailed picture of the stupendous waterworks that supported Khmer civilization. Angkor's kings and their lavish works depended on huge rice yields sustained throughout the year; these in turn required enormous amounts of water and extensive irrigation systems. A radar shot from the space shuttle *Endeavour* in 1994 revealed segments of the so-called Great North Canal that moved water from the northern hills to two reservoirs (see sidebar). Today, an international team of researchers, headed by Roland Fletcher, Christophe Pottier, and others, is using more NASA radar images, state-of-the-art GPS technology, and even an ultralight aircraft, to map Angkor's huge artificial landscape of dwellings and water tanks, once connected by small roads and canals, which extends over about 386 square miles (1,000 square kilometers). Three great reservoirs, or barays, stilled and stored the waters of three rivers, then diverted it, as required, for the ceremonial pools and storage facilities of the great temples, and also through canals to irrigated fields in the southern half of Angkor, and also as a way of stemming floodwaters. The reservoirs would have supplied enough water to support between 100,000 and 200,000 people, out of an estimated population of about 750,000. Most Khmer

Archaeology from Space

Satellite photography and related imagery provide fascinating perspectives on the past from space. These technologies are especially useful for studying ancient land use and to locate specific features like fortifications, long-abandoned irrigation canals, and reservoirs. A number of different instruments can scan the electromagnetic spectrum emitted from the earth's surface. The thermal infrared multispectral scanner (TIMS) uses a six-channel scanner to measure thermal radiation on the ground with great accuracy. The temperatures of soils and other sediments are of course invisible to the human eye, but so easily recorded by TIMS that minor differences in soil texture and moisture have revealed ancient field systems and roads masked by vegetation in the Maya lowlands. The images, with their overhead view, provided vital information on the intensive farming of the rainforest environment by Maya farmers, which could never be detected on the ground.

Synthetic aperture radar (SAR) beams energy waves to the earth's surface, which provide returning signals that reflect long-forgotten features on the ground. SAR is especially useful for detecting linear and geometric features, such as the canals, reservoirs, and roads that surround Angkor Wat detected by the space shuttle *Columbia* in 1981. Satellite instruments can even be programmed to detect specific kinds of features such as canals. Over the Sahara Desert, *Columbia* traced long-hidden watercourses and buried valleys in the heart of the desert. When geologists investigated on the ground and dug into the hidden defiles, they were astounded to recover 200,000-year-old stone axes dating from a time when the Sahara was better watered than today.

Satellite imagery from space is expensive, but it provides unique, and often unexpected, perspectives on human exploitation of the landscape on a much larger scale than one might expect. The Ances-

tral Pueblo of Chaco Canyon, New Mexico, built an elaborate road network that converged on the canyon. No one realized how extensive the road system was until satellite photographs revealed more than 370 miles (600 kilometers) of incomplete road segments, largely invisible except from space. The purpose of this road network remains a complete mystery, for the roads were not "highways" in the western sense, with clear destinations. They probably had a role in defining a now forgotten symbolic landscape.

lived off independent rice paddies that captured water during the monsoons. But the barays filled one important function: they would store water that could be used in poor harvest years.

The Angkorian empire was in decline by the fifteenth century, and urban Angkor itself was abandoned by the end of the sixteenth. Quite why remains the subject of vigorous controversy. Did trade routes shift away from Angkor? Had the temple-building monarchs bankrupted an exhausted kingdom? Did the rising influence of Buddhism eclipse the influence of Hindu god-kings? Most likely, a precipitous decline in crop yields was a decisive factor. Rising silt levels in canals during the dry season may have choked water supplies, the result of soil erosion resulting from widespread deforestation. Today, and as a result, the main river is now more than 16 feet (5 meters) below the ancient ground surface. The Greater Angkor Project, led by Roland Fletcher, is studying the canals and spillways. Fletcher argues that Angkor's water management system became ever more complex as the sprawling civilization grew. Over many generations, the system became too elaborate, so huge that it could never be completely overhauled to compensate for the extremes of flood and drought that are inevitable in a monsoon zone. The Khmer created a fragile, totally artificial environment that was ultimately as unsustainable as it was magnificent.

As long as cool and La Niña–like conditions prevailed, summer monsoons brought good rains, but, as the Medieval Warm Period gave way to the Little Ice Age, climatic conditions may have become more volatile, with a higher incidence of El Niños and droughts. And the overtaxed water management system of Angkor would have been unable to supply the insatiable demands of god-kings. The collapse would not have come at once, but as a slow death when the people gradually dispersed into smaller settlements. The architectural masterpieces of Angkor were left as desolate as the empty Maya water mountains of the Yucatán.

CHAPTER 12

China's Sorrow

Disorder, like a swelling flood, spreads over the whole empire, and who is he that will change its state for you?

—Confucius, *Analects*[1]

THE HUANG HE BASIN, NORTHERN China, late winter, A.D. 950. The wind cuts through one's clothing, chilling one to the bone and gumming up one's eyes. Oblivious to the cold and the fine dust, the farmers toil over their arid plots of land, thick cloths sheltering their faces. They turn over the soil, breaking up sods of earth hour after hour with never a break. The men move slowly, deliberately, stoically, as if knowing that their efforts are in vain. Last year's millet harvest was far less than usual after a hot, unusually arid summer. People have been dying of hunger and dysentery for months, but the winds never relent, the skies are gray, the endless dust accumulates pitilessly on the freshly turned, arid soil. One of the farmers looks wearily up at the gloom overhead, looking vainly for even a hint of spring rain. There is more hunger ahead.

They call the Huang He (Yellow River) "China's Sorrow" because it has killed millions of people with its sudden floods and lengthy droughts. Few rivers are more prone to disaster than the Sorrow, at

3,395 miles (5,464 kilometers) China's second longest river after the Yangtze. The Huang He rises in the Kunlun Mountains south of the Gobi Desert, flows through some deep gorges, then across the Ordos Desert before emerging into a huge drainage basin carved out of extensive plains covered with the fine, windblown dust known to geologists as loess.[2] Here the great river picks up a heavy load of fine silt which turns its water a distinctive yellow. Eighty-seven miles (150 kilometers) from the mazelike channels of the mouth, the silt load exceeds that of every river in the world except the Ganges-Brahmaputra and the Amazon. With irregular monsoon rainfall and savage droughts, the 334,000 square miles (865,000 square kilometers) of the Huang He basin have been a crucible for human misery for more than seven thousand years. Here, global climatic forces helped decide the fate of medieval Chinese societies.

Map showing locations and peoples mentioned in this chapter. Some minor places are omitted for clarity.

In northern China, the monsoon and the forces that drove it shape the climate of the warm centuries. As always, the climatic record of the warm centuries comes, for the most part, from proxies. The documentary record is long. For more than a thousand years, Japanese and Korean officials have recorded the date of cherry trees' spring flowering, a historical record whose duration rivals the longest from Europe. By combining archives such as these with proxies, Chinese climatologists have developed a winter temperature curve for eastern China, which shows that readings were above the long-term mean from A.D. 950 to 1300.[3] The Medieval Warm Period was reality here. But, as always in East Asia, the major climatic player during these four centuries was the monsoon, nurtured in the Pacific Hot Pool.

The East Asian monsoon has close links to the Southern Oscillation, El Niños, and La Niñas. A long-term research program led by Wang Shao-wu at Beijing University has shown that when an El Niño warms the eastern tropical Pacific in winter, the subtropical high intensifies and shifts westward the following summer.[4] This movement blocks the

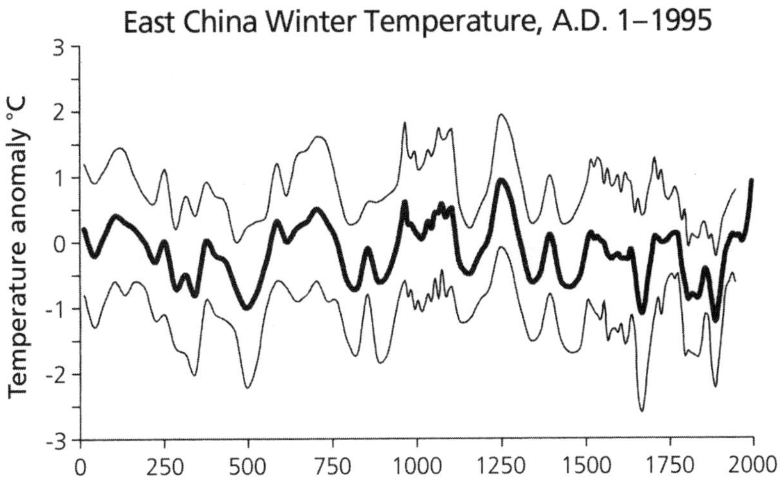

The dark line is the estimate, with one standard deviation shown on either side. The chart is based mainly on historical records.

summer monsoon from moving as far north as usual, so the Huang He basin suffers from sparse rainfall or drought. Ever since 1870, weather observations from stations in northern China have shown a connection between El Niño events and widespread drought. The East Asian monsoon stalls over the middle and lower Yangtze. Heavy rain falls there in June and July, while the north suffers from intense drought. When the Southern Oscillation swings and a cool, dry La Niña sets in over the Pacific, the subtropical high no longer blocks the northward movement of the monsoon, so summer rain falls in the north and often brings widespread flooding, with drier conditions in the south. The climatic contrasts caused by the ENSO-connected monsoon are such that there were effectively two Chinas for over three thousand years, from before 1000 B.C. In the south, the visitor was struck by the bustling commerce of the Yangtze Valley. In the north, poverty-stricken subsistence farmers wrestled with uncertain rainfall. (One should caution that the links between ENSO and monsoons and between El Niños, La Niñas, and climatic conditions in China are, to put it mildly, complex and still little understood.)

Forty-five percent of China's population lives in the provinces north of the Yangtze, in regions where rainfall can vary by 30 percent from year to year. The Huang He Valley receives 70 percent of its 19.6-inch (500-millimeter) annual rainfall between May and September, during the hot summer. Severe, dry winters with blowing snow add another challenge for northern farmers. For thousands of years, they have harvested wheat in June, then millet and sorghum in September. If the irregular spring rains fail, there is a poor wheat crop. If the summer monsoon falters, then there is no harvest for the entire year. The critical month is June, with poor rains a high probability. For instance, in modern times, Beijing has received deficient June rainfall during twenty-one of fifty-five years. Five of those years saw virtually no rain at all. Uncertain rainfall and violent climatic extremes made the Huang He a risky environment for cereal farmers even in good years. Everyone lives at the mercy of the monsoon.

A thousand years ago, temperatures were somewhat warmer in the north, where the Medieval Warm Period was in full swing. The American geographer George Cressey wrote in 1934: "With more adequate rainfall, it [the loess soil region] might form one of the most productive soils in the world."[5] For details, we have to rely on climatic proxies, which are still few and far between.

One such proxy comes from the other end of China, from Lake Huguangyan on the low-lying Leizhou Peninsula in the tropical south.[6] The closed lake lies in a region where 90 percent of the annual rainfall falls between April and October, the amount being determined by the position and intensity of the subtropical high in the western Pacific. The carbonate concentrations in the lake sediments vary considerably over time, probably as a result of changing evaporation rates and rainfall fluctuations. High carbonate levels signifying dry conditions occur at Huguangyan between A.D. 880 to 1260, which coincides with a widespread low-moisture index recorded in eastern China and major lake level changes throughout the entire country.

The Chinese droughts even have a global connection. Thousands of miles to the northeast, the Ohio State University climatologist Lonnie Thompson, well known for his work on Andean glaciers (chapter 9), drilled a series of cores deep into the Guliya ice cap on the Qinghai-Tibetan Plateau, where glaciers cover an area of 22,000 square miles (57,000 square kilometers).[7] Guliya is in the western Kunlun Mountains. Four hundred and thirty-three feet (132 meters) of one core cover the past two thousand years and reveal an extended dry period between A.D. 1075 and 1375, followed by a nearly four-century wetter interval. Thompson was fascinated to discover that the Guliya drought coincided almost exactly with a major drought of the twelfth and thirteenth centuries 12,400 miles (20,000 kilometers) away in the Quelccaya ice cap of the southern Andes.

While Guliya links South America and Asia, cores sunk into Lake Huguangyan in coastal southeastern China provide a high-resolution record of climatic change in the south and another global connection. The sequence goes back over sixteen thousand years, measured by studying changing magnetic properties and titanium content, the latter a record of sediment deposition, in the lake deposits as proxies that measure the strength of dry winter winds.[8] During periods of warming in the northern hemisphere, the summer monsoon was stronger and the winter monsoon was weak. When the ITCZ moved south, as it does during El Niño years, the summer monsoon was weak and there was less rainfall. The Huguangyan cores show a general shift toward drier, colder climate in about A.D. 750 to A.D. 900, with a series of three multiyear droughts within that generally dry period.

Remarkably, these droughts coincide with the dry cycles recorded in the Cariaco basin deep-sea cores off Venezuela, described in chapter 8. The Cariaco records chronicle multiyear droughts that began as early as A.D. 760 and recurred at about fifty-year intervals: 760, 820, 860, and 910. The Cariaco droughts occurred when the ITCZ moved southward, just as they did in China, at a time of prolonged La Niña conditions in the tenth-century eastern Pacific, known to us from the Palmyra Island corals. Cool La Niña conditions in the tropical Pacific usually mean that heavy monsoon rains fall in northern China, while the south is dry, exactly the pattern one sees in the few proxies there.

For the northern Chinese, the Medieval Warm Period may have been warmer, but it was a time of violent climatic swings nurtured thousands of miles away that brought either lengthy dry cycles or torrential rainfall that inundated thousands of acres of the Huang He basin.

MEDIEVAL AGRICULTURE IN the Huang He basin lived with sudden climatic twists and turns, and with the sins of earlier farmers. Loess forms a fine, soft-textured earth that is both homogeneous and porous, making it easy to cultivate with simple ox-shoulder-bone shovels and digging sticks. Cereal farming began here at least 7,500 years ago, in the midst of a verdant, forested landscape, where summer monsoon rains and a dry winter nourished small agricultural settlements. Rainfall was higher than it is today; droughts were but a sporadic problem for over three thousand years. Archaeological surveys have located dozens of prosperous villages across a densely populated landscape dating to before 2000 B.C.[9] Then, suddenly, the number of archaeological sites shrinks dramatically between 2060 and 1600 B.C. as more arid conditions settled over the region. Population densities fell rapidly, less fertile areas were abandoned, and people retreated from higher elevations. Plant and tree pollens from the extensive clay silt layers in the valleys show that forests gave way to grasses and shrubs as a result of prolonged drought and perhaps some human clearance. For the next two thousand years, dry conditions persisted, making agriculture at best a marginal activity. The survivors of what must have been a major catastrophe turned to sheepherding in the now barren landscape. Only about two thousand years ago did wetter conditions once again prevail and agriculture on any scale resume.

The climate shift of four thousand years ago was an event on a millennial scale, but northern China has never been very wet, so what should have been a paradise for cereal farmers never realized its full potential. They were at the mercy of unpredictable summer monsoon rains and a capricious environment that drought, human deforestation, and sheep grazing had modified fundamentally. As farming populations rose during the second millennium and the primordial Shang civilization flourished along the Huang He and its Wei River tributary, much of the remaining forest vanished, with disastrous consequences. Heavy summer rains washed cultivated hillsides into the river and left a scarred and eroded landscape behind them. Over the centuries, river-borne silt accumulated rapidly, thereby accentuating summer flooding on the plain. The river channel itself was never stationary, shifting suddenly and arbitrarily. Even when the rains were plentiful, this was a high-risk neighborhood for village farmers, despite small-scale irrigation canals and numerous wells built and maintained by villages and households. Crop yields were low, reflected in thousands of years of a political economy that was based on small-scale peasant farming by villagers living in compact settlements located on higher ground.

The loess highlands and delta of the Yellow River are unique, not for the fertility of their soils, but because of the frequency of disastrous floods and especially droughts. Historical floods and droughts make the point, for conditions had changed little a thousand years later, in the late nineteenth and early twentieth centuries. In A.D. 595, the emperor Yang-Kien was forced to move his court from Xi'an to Henan, as there was not enough food for his court. Nineteenth-century droughts were of epochal proportions and accentuated by political unrest. In 1877–79, a third of the population of Shaanxi Province died of hunger and famine-related diseases.

The years 1897–1901 brought another savage drought. More than 2 million out of a total population of about 8.5 million died. An American journalist, Francis Nichols, traveled to Xi'an, the ancient capital of China, as a special "famine commissioner" for the *Christian Herald* of New York in 1901.[10] A prolonged drought lasting three years, combined with Shaanxi's isolation from the coast and with the fact that the region is surrounded by mountains, made it virtually impossible to bring food in from outside. The loess soils had turned to a "dry, white powder, in which crops parch and wither and die." No rain fell from the summer of 1898 until May 1901. The farmers' small food reserves were soon exhausted. Wells and rivers dried up. The country became a vast desert. The price of a bushel of wheat rose fifteenfold in a few weeks.

As their fields dried out, so thousands of farmers moved into Xi'an, 300,000 of them during 1900–01 alone. The governor forbade them to come within the city walls, so they camped in caves dug into riverbanks and in the fields. They ate coarse grass and weeds. Nichols visited "grim, blackened caves" around Xi'an, nearly all of them empty, their inhabitants long dead. Dysentery and cholera followed in the wake of famine. At the height of the disaster, the governor's officials were burying more than six hundred corpses a day. Nichols reports that cannibalism became inevitable. "A horrible kind of meat ball, made from the bodies of human beings who had died of hunger, became a staple article of food, that was sold for the equivalent of about four American cents a pound."[11]

The authorities received funds from Beijing and set up soup kitchens, but the main problem was a lack of food locally with which to feed the starving. Entire families subsisted on cats and dogs, also horse meat, then slowly starved to death. Nichols traveled through the countryside. "Every quarter of a mile, a village rose out of the white, treeless desert, which stretched away to the north, east and west like a treeless ocean. The vast plain was silent. . . . No farmers were in the fields. . . . The plain was silent because its inhabitants were dead."[12]

Floods also took their toll. Heavy monsoon rains pummeled the Huang He basin in early summer 1898. The river flooded and overflowed its bank in Shouzhang, then farther downstream. Over two thousand villages and 2,972 square miles (7,700 square kilometers) of farmland vanished under water. Millions fled, many stranded on dikes, where they lived off willow leaves, wheat gleanings, and cotton seed. Tens of thousands perished. Another catastrophic inundation, in August 1931, killed an estimated 3.7 million people.

These droughts and floods occurred when the Huang He was still effectively a preindustrial farming area. Much of the destruction resulted from inept irrigation works, poor water management, and corruption, but the potential for disaster was always there. It is even greater today, for the Huang He is an extreme example of hydrological crisis. The river is almost at the point where it cannot support any more water exploitation. In the early 1900s the low-flow period in the river was about forty days. Today, the low-flow period lasts two hundred days, which places severe stress on the more than 100 million people living in the Huang He basin and on their ability to grow crops, quite apart from reducing freshwater species and habitats.

A.D. 850, AT the Chinese court. The long line of northern tribesmen, headed by a brightly caparisoned Mongolian khan, rides along a beaten-earth road to the emperor's palace. Solemn Chinese officials surround the visiting leader. They have met the envoys at the border and accompanied them to the capital. The heavily laden column dismounts. The goods are unloaded and stored, as court officials coach the visitors in the proper etiquette for their appearance. When the day of the audience comes, the khan and his leading henchmen are escorted into the imperial presence. With detached care, they perform the expected rituals and kowtow before the emperor in a symbolic gesture of inferiority. The visitors are permitted a brief conversation with the emperor; hides and horses, some falcons are presented as gifts. In return they receive lavish gifts from their royal host. The audience soon ends. The closely supervised envoys are now permitted to trade with Chinese merchants for three to five days.[13]

The Chinese emperor, who had a Mandate of Heaven to rule his domains, set an example of orderly government and society that would encourage foreigners to be "transformed." His virtuous actions were thought to offer an irresistible attraction to the "barbarians," who dwelled outside the realm of Chinese civilization.

Such was the idealized Chinese vision of a self-sufficient empire indifferent to foreign lands. Reality was, of course, far more complex and rooted in centuries of complex and often violent interaction between the Chinese of the settled lands and the nomadic peoples of the northern grasslands.

When the irregular dry cycles recorded in lake and ice cores descended over East Asia during the ninth century, T'ang emperors held sway over China.[14] This three-century dynasty (A.D. 618–907) was a high point in Chinese civilization. Early T'ang lords, based at Ch'ang-an (present-day Xi'an), acquired their empire by conquest and maintained trading contacts by land and sea with India and southwestern Asia. The Silk Road across Eurasia enjoyed great prosperity. Thousands of foreigners lived in Ch'ang-an, at the time one of the great cosmopolitan cities of the world. Kashmir and Nepal, Vietnam, Japan, and Korea paid tribute to the T'ang, while nomadic Eurasian tribes called the emperor Tian Kehan ("Celestial Kaghan"). The T'ang emperors were remarkable for their religious tolerance during three centuries when Buddhism became part of Chinese culture, printing was invented, and both literature and art enjoyed a golden age. T'ang rule thrived because of a system of government that relied on trained career officials who had no territorial base or local loyalties. Many of them were scholar-bureaucrats, who acted as intermediaries between government and commoners.

By the middle of the eighth century, T'ang power was eroding, following a defeat by the Abbasid caliphs at Talas somewhere in Kazakhstan. They were struggling for control of key trade routes in central Asia. The T'ang were eventually driven out of central Asia, and China did not regain power there until Mongolian times. During the late ninth century, a series of persistent rebellions by powerful provincial lords weakened the central authority of the government. The last emperor was deposed in A.D. 907.

But the most powerful enemy of the T'ang dynasty may have been cold, dry conditions and strong winter monsoons with less summer rainfall. If later history is any guide, crop failures and hunger fostered social disorder and rebellion. The T'ang may have been powerless to maintain central control in a situation of persistent and widespread drought in the loess lands around Ch'ang-an, just as Maya civilization on the other side of the world collapsed over a wide area in a time of volatile droughts. The task of supplying food to millions of people may simply have been beyond the capability of the government of the day. As Francis Nicholson saw, the same sort of crisis brought this part of China to its knees even at the turn of the twentieth century.

After 900, China fragmented into five northern dynasties and ten southern kingdoms. With dizzying rapidity, ambitious lords rose to power and were just as promptly dethroned. Given the political vacuum, the political situation would have been difficult at the best of times. But, if the climatological sequences are to be believed, northern China was also plagued by unusually dry conditions, and by prolonged, often serious droughts. The crop losses and resulting famines must have added complex variables to an already volatile frontier between lands inhabited by settled farmers and nomadic herders.

We tend to think of the northern frontiers of China as a rigid entity, defined by the Great Wall. Reality was very different. The T'ang never presided over a clearly demarcated northern frontier, merely a scattering of fortresses and military colonies and some fortified border prefectures. (The current Great Wall was built during the Ming dynasty, after 1449.) They had believed in defense in depth, backed by powerful armies in the provinces far from the border. They also maintained a complex set of agreements with the tribes in the border districts, whereby the tribal leaders maintained their independence, but were given Chinese titles and ranks. Large numbers of prominent tribespeople attended the T'ang court, but the constant intercourse did not make Han Chinese out of them. Instead, they acquired a firsthand knowledge of the court and of Chinese institutions and administrative methods, which was to stand them in good stead in later times. For centuries, the border was partly an ecological boundary between the landscape of settled farmers and environments where only herders could flourish. But the frontier was also a multilayered region where Han Chinese and nomad lived amicably alongside one another, while cultural and ethnic identities remained distinct. When the T'ang dynasty faltered, the borderlands still remained a permeable frontier, but now became a region where military leaders held sway.

The droughts of the warmer centuries lasted for long periods of time in western North America and the Andes, and, if the Guliya and Huguangyan cores are any guide, so did those in East Asia. The droughts were not continuous, but cyclical, which would have had dangerous shock effects in the loess lands where the northern borderlands lay. When a sudden wet year followed a long drought cycle, floods would have inundated the arid fields and disused irrigation works in short order. The centuries of the Medieval Warm Period were climatically extremely volatile in this region of dramatic rainfall shifts, perhaps even more so than almost anywhere else on earth. The vagaries of drought and flood must have rippled through the realms of politics and war, for both farmer and nomad lived at the subsistence level and at the mercy of the climate, whatever the deeds of great lords and warring armies.

THE EXTREME CYCLES of medieval climate also affected the complex relationships between the nomadic peoples of eastern Eurasia and those who dwelled on the settled lands. The most powerful of these tribal groups were the Khitan, herders and horse people whose origins went back deep into the past.[15] Like those of other nomadic peoples, their lives were governed in part by rainfall on the steppe, by the gyrations of the desert pump. In periods of drought, they pressed southward into better-watered and more settled lands. After the 840s, and as the T'ang hold on power weakened during drier cycles, the Khitan defeated their tribal neighbors, then turned their attention to the powerful states to the south. At first the Khitan contented themselves with raids and temporary incursions into more settled lands, after which they withdrew northward again. To what extent these movements were the consequence of drought on the steppe, we do not know, but, judging from centuries of nomad history, many of the most serious incursions certainly occurred during dry years when grazing was in short supply.

The collapse of T'ang power and increasing rivalry between different warlords operating in the borderlands led the Khitan to unite. With the accession of A-pao-chi as Great Khan in A.D. 906–07, the Khitan embarked on ambitious campaigns of conquest. Within twenty years, they had become masters of the nomadic peoples of Mongolia and Manchuria. Theirs was a well-organized kingdom, with cities for Chinese from the border regions, a diversity of industries and areas of settled farming, and a dual form of organization that accommodated both the Chinese and nomadic ways of life. The pattern of nomadic life was changing, as farmer and herder became increasingly interdependent, a useful form of insurance in climatically volatile times.

A-pao-chi died in 926, to be followed by the Liao, Hsia, and Chin states during a period of constant warfare and seething rivalries. But behind all these political and military events and a pastiche of rulers lay the harsh economic realities of subsistence agriculture on the settled lands. We know from Chin records that the state produced about 90 million *shih* (a *shih* is about 125.5 pints [59.4 liters]) of millet and rice annually. A tenth of that went to the government as land tax. The average grain consumption of an individual was about 6 *shih* annually, so a year of good rainfall produced just enough grain to feed the population adequately. But an average year left no surplus to build up reserves for distribution during droughts. The food supply was never secure in the north, as it was in the southern Sung kingdom in the rich environment of the Yangtze Valley, which produced as many as two rice harvests a year.[16]

Chin rulers were well aware of the precarious food situation and attempted to increase the acreage of land under cultivation by fostering irrigation works. They also attempted to increase crop yields by terracing hillsides. But both these measures had unforeseen consequences, especially terracing, which led to inexorable deforestation and rapid soil erosion. The latter had a particularly serious effect in the Huang He basin.

Agricultural production was precarious, even during good years, so the drought cycles of the warm centuries must have had a serious impact on political events in the north. Contemporary records do not dwell on droughts and other natural disasters—hardly surprising, for the peasantry were illiterate, anonymous, and almost a "background noise" to the goals of warlords, emperors, and ambitious officials. But the vagaries of the summer monsoon made it imperative that northern states import rice from their southern neighbors, the Sung. The need had been there for centuries. Overland transport was slow and unreliable, so the logical way to transport grain was by water. Coastal routes were unreliable and dangerous because of pirates and storms; inland waterways linking the Yangtze and Huang He basin were the best solution, although punishingly expensive.

Efforts to construct a waterway had begun as early as 486 B.C. The Sui rulers of the late sixth and early seventh centuries A.D. linked earlier sections, joining the rich agricultural regions of the lower Yangtze with their western capital at Luayang. A patchwork of lakes and canals became the Grand Canal, Da Yun He, the longest artificial waterway in the world, far longer than Suez or Panama. By the tenth century, the system boasted locks, feeder lakes, and lateral canals. At its peak during the fifteenth and sixteenth centuries, the canal system extended over 1,553 miles (2,500 kilometers), ran through 24 locks and under about 60 bridges, and carried about 441,000 tons (400,000 tonnes) of grain annually.

For centuries, northern China and its loess lands were dependent on critical food supplies from the south. Not even the most organized preindustrial state and effective administration could overcome the prolonged droughts and sudden floods that regularly devastated food supplies and those who produced them. The north was profoundly vulnerable a thousand years ago. The Huang He basin is even more vulnerable to catastrophe today.

CHAPTER 13

The Silent Elephant

I have seen a herd of elephants traveling through dense native forest . . . pacing along as if they had an appointment at the end of the world.

—Isak Dinesen, *Out of Africa*[1]

AN ARID LANDSCAPE IMPRINTS ITSELF on your mind. I remember as if they were yesterday November days of over forty years ago in central Africa. Every morning, after weeks of intense heat, a brazen sun rose from a dusty horizon, not a cloud in the heavens. The temperature climbed as the shadows shortened. A dusty blue parabola of cloudless sky reflected the heat radiating from the parched earth. Occasional gusts of wind propelled williwaws across the grassland. Humans and beasts alike sought the mocking illusion of cool shade under trees or roof eaves. Cattle stood motionless, heads drooping, waiting patiently for the cool of evening. I watched farmers gaze stoically at the maize withering in the cracked fields, planted a month before when a heavy shower brought a promise of more rain. As the sun descended, you deluded yourself that the temperature was falling. But it was still 87 degrees F (30.5 degrees C) at midnight. Hunger was not far over the horizon.

By all accounts, the droughts of the next century will be infinitely worse than this one.

When I started researching this book, I was expecting to find widespread evidence of increasing temperatures a thousand years ago, of startling changes in agricultural practices, and of ocean voyaging and prosperity in environments basking in unaccustomed warmth. The earlier chapters indeed found me exploring a bustling Europe with bountiful harvests. I followed Norse voyagers across the North Atlantic as they developed fleeting contacts with the Inuit of the far north. So far so good—but when I traveled to the Eurasian steppes, to the West African Sahel, and to the Americas, I encountered major and prolonged droughts that changed history.

I stress the word "prolonged." The dry spells of a thousand years ago spanned not years, but generations. The medieval droughts in California's Sierra Nevada lasted for decades, far longer than those of modern times. A long drought cycle lasting half a century triggered major adjustments in Ancestral Pueblo life in the Southwest, as we have seen. Drought settled over Nebraska and the Plains.

The Southwest has always been arid, but is not the only part of North America to suffer drought. Pollen cores from the Piedmont Marsh of the lower Hudson River Valley on the East Coast chronicle drought conditions between A.D. 800 and 1300, times when the estuary became saltier. If similar drought conditions were to prevail in the same area today, the water supplies of millions of people would be endangered, among them the citizens of nearby Poughkeepsie, New York, which draws its water supplies from the Hudson, as do other suburban towns in the area.[2]

Far to the south, in Central America, great Maya cities tottered under medieval drought while Andean civilizations wilted in the face of an evaporating Lake Titicaca and faltering runoff in coastal river valleys. Looking at the global picture, it is tempting to rename the Medieval Warm Period the Medieval Drought Period.

THE REVOLUTION IN climatology began in earnest about thirty years ago, when techniques for deducing the climatic record from proxies, such as deep-sea cores, ice borings, coral, and tree rings, entered the scientific mainstream. Satellite observations and computer modeling joined the meteorological armory during an explosion of research into El Niños and the Southern Oscillation. Since the 1980s, humanly caused global warming has engaged the attention of climatologists in the face of well-documented, virtually continuous warming since 1860. Suddenly, the climate changes of the past two thousand years have assumed great importance in the public arena as anthropogenic warming has become a scientific reality and a major political issue. For this we must thank not only Al Gore and his documentary on global warming, but also a growing public consciousness that rising temperatures, more extreme climatic events, and higher sea levels are facts of life for humankind's immediate future.[3]

Almost immediately, the Medieval Warm Period assumed great importance to the debate over warming in many people's minds.

"We've been through this before," both supporters and debunkers of global warming cried. Claims and counterclaims rocketed across auditoria from Tokyo to Scandinavia, as scientists, journalists, and activists argued over whether the Medieval Warm Period was warmer than the rapidly heating world of today. As new tree-ring sequences and other evidence have became available, so the debate about Warm Period temperatures has intensified, with no end in sight.

The Medieval Warm Period is still a shadowy entity, but we know a great deal more about it than we did in Hubert Lamb's day. A growing number of sources tell us that there was never long-lasting medieval warmth, but that between 1000 and 1200, temperatures were a few degrees warmer *in some parts of the world*, notably parts of China, Europe, and western North America.

Today's preoccupation with medieval warmth is entirely understandable in a time of uncontrolled anthropogenic warming, with all the harrowing threats of melting Greenland ice sheets, rising sea levels, and increased storminess.

What would happen if the melting Greenland ice sheet partially shut down the Gulf Stream? Would Europe be plunged into a near–Ice Age, as indeed happened some twelve thousand years ago during the climatic episode known as the Younger Dryas, named after a polar flower?

What would happen to the Low Countries and to some Pacific atolls if sea levels rose as much as a foot (0.3 meters) or more by century's end as a result of partially melted ice sheets?

These are perfectly legitimate concerns, which will require concerted political will to solve in coming generations. But our preoccupation with heat and rising sea levels ignores an even greater threat: drought. Why this surprising neglect? Undoubtedly the devastation of the Southeast Asian tsunami in 2004 and Hurricane Katrina the following year reinforced fears about extreme weather events and flooding in particular. But these two events, coming in two of the warmest years since the Ice Age, seem to have delivered a message that warmer centuries mean more rain, not less. Then there's another reality: most, though not all, of the people likely to be affected by severe drought in the future live in the developing world, and we in the United States are still much preoccupied with the flooding brought by Katrina.

I COULD HEAR the Zambezi River riffling in the rocky shallows, in the distant background the unceasing roar of Mosi-oa-Tunya, "The Smoke That Thunders," Victoria Falls. Dense bush pressed on the clearing, trees arching overhead, dry leaves rustling softly in the afternoon heat. I was completely alone—or so I thought. Then I heard the sound of trampling and crashing branches: I realized with horror that I had walked into the midst of a small herd of elephants. The great beasts were invisible but close by, seemingly unaware of my presence. I tiptoed back the way I had come until I emerged from the trees. As I reached the Zambezi, I looked back. A huge bull elephant flapped his ears at me, feet firmly set in the shallows. He watched closely, unmoving, as I beat a careful retreat.

Elephants can tread delicately when they wish and can easily become invisible until it is too late to avoid them.

When the novelist George Orwell, of *1984* fame, was a police officer in Burma in the 1930s, he was confronted with a berserk elephant in a bazaar. At a distance, "peacefully eating, the elephant looked no more dangerous than a cow." But the beast had killed a man, and "a mad elephant had to be killed like a mad dog."[4] Orwell was struck by the violent contrasts in his by now seemingly placid prey. And so it is with drought. As my research progressed away from Europe, I realized that drought was the hidden villain of the Medieval Warm Period. Prolonged aridity was the silent elephant in the climatic room, and the unpredictable swings of the Southern Oscillation were what brought the beast through the door.

A surge in ENSO research over the past twenty years has revealed that El Niños, and their sister La Niñas, are not merely local phenomena, but, next to the passage of the seasons, among the most powerful factors in global climate change. Major ENSO events bring heavy rainfall and floods to the Peruvian coast and torrential precipitation to California; they reduce the frequency of tropical storms and hurricanes in the Atlantic. They also bring severe drought to Southeast Asia and Australia, to Central America and northeast Brazil, and to parts of tropical Africa. The less conspicuous, and often longer-lasting, La Niña can be just as destructive, especially in its ability to nurture drought over large tracts of the world—as was the case during the Medieval Warm Period, when the cool, dry sister of El Niño persisted for years on end.

WHETHER THE MEDIEVAL Warm Period was warmer than today, and why, is still a matter of much debate. Our current warming has not lasted nearly as long as the period studied in this book. It is, however, a steady and well-documented trend, with no downward curve in sight. And unlike the situation a millennium ago, humans are numerous enough, and our outputs profuse enough, to push the trend further and faster. What is not debatable is that if we reenact the climate history of a millennium ago—let alone see the earth get even warmer—we will see how vulnerable humans are to the forces of their environments.

But if you look at the warm centuries with a global perspective, the wide incidence of drought is truly striking and offers a sobering message about tomorrow's world. Prolonged aridity was widespread in medieval times and killed enormous numbers of people. Evidence is mounting that drought is the silent and insidious killer associated with global warming. The casualty figures are mind numbing. About 11 million people between Kenya, Somalia, Ethiopia, and Eritrea were in serious danger of starvation as a result of multiyear droughts in 2006. The International Institute of Tropical Agriculture in Nigeria estimates that by 2010 around 300 million people in sub-Saharan Africa, nearly a third of the population, will suffer from malnutrition because of intensifying drought.[5] (Relatively few people die of hunger during a drought. They perish from epidemics of dysentery and other diseases spread by poor living conditions. For instance, 1.6 million children a year die today because of a lack of access to good sanitation and clean drinking water.)

The long-term future is even more alarming. A study by Britain's authoritative Hadley Centre for Climate Change documents a 25 percent increase in global drought during the 1990s, which produced well-documented population losses.[6] The Hadley's computer models of future aridity resulting from the impacts of greenhouse gas emissions are truly frightening. At present, extreme drought affects 3 percent of the earth's surface. The figure could rise as high as 30 percent if warming continues, with 40 percent suffering from severe droughts, up from the current figure of 8 percent. Fifty percent of the world's land would experience moderate drought, up from the present 25 percent. Then the center ran the model without factoring in the impact of greenhouse gases, which they assumed were the temperature change villains. The results implied that future changes in drought without anthropogenic warming would be very small indeed.

In human terms, the United Nations Environment Program reports that 450 million people in twenty-nine countries currently suffer from water shortages.[7] By 2025, an estimated 2.8 *billion* of us will live in areas with increasingly scarce water resources. Twenty percent of the world's population currently lacks access to safe, clean drinking water.

Contaminated water supplies are a worse killer than AIDS in tropical Africa. If the projected drought conditions transpire, future casualties will rise dramatically. The greatest impact of intensifying drought would be on people already living in arid and semiarid lands—about a billion of us in more than 110 countries around the world. And those who would be hit hardest are subsistence farmers, especially in tropical Africa. Seventy percent of all employment in Africa is in small-scale farming, and completely dependent on rainfall.

The number of food emergencies in Africa each year has already almost tripled since the 1980s, with one in three people across sub-Saharan Africa being malnourished. The Nigerian institute's projection for 2010 is just the beginning. Future drought-related catastrophes will make these preliminaries seem trivial and could affect more than half of tropical Africa's population.

Peru offers another frightening example. The Cordillera Blanca, the largest glacier chain in the tropics, is melting fast because of rising temperatures. The Quelccaya ice cap in southern Peru, a crown jewel of

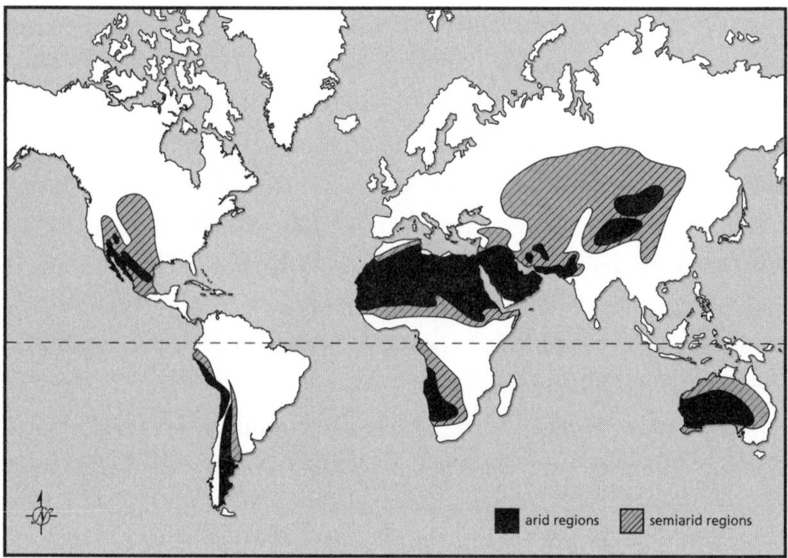

Arid and semiarid regions of the world today.

climatic data, is retreating about 197 feet (60 meters) a year, three times faster than in the 1960s. Taken as a whole, the Peruvian Andes have lost at least 22 percent of their glacier area since 1970. Two thirds of Peru's 27 million people live on the coast, where only 2 percent of the country's water supply is to be found. A thousand years ago, with many fewer people to feed, the lords of Chimor could adapt their irrigation strategies to prolonged droughts. Their modern successors, living in crowded cities, shantytowns, and an increasingly congested rural landscape hemmed in by desert, cannot do so.

Droughts are expensive in human terms and also carry a high economic price. The notorious Dust Bowl droughts of 1934–40 over the Great Plains scarred an entire generation. Three and a half million people fled the land. Many suffered from typhoid and other diseases, also long-term health effects like higher risks of cancer and heart disease, but exact casualty figures are not available. The midwestern drought of 1950 to 1956 brought extreme heat, ruined numerous farmers, and reduced crop yields in some areas by as much as 50 percent. The 1987–89 drought covered 36 percent of the United States, less than 70 percent of the area affected by the Dust Bowl, but with an estimated cost of $39 billion, which makes it one of the most expensive natural disasters in American history. The dry conditions fanned huge wildfires in the West and caused serious navigational problems due to drought-related shoaling in the upper Mississippi basin. As a comparison, and to give some insight into the potential costs of really huge future climatic disasters, 2005's Hurricane Katrina has cost $81 billion so far, and that figure is rising.

History tells us that droughts have long wreaked havoc, especially in the tropics. At desert margins and in semiarid environments, the pump effect comes into play as faltering rainfall pushes animals and people out to better-watered margins. In a telling analysis of nineteenth-century droughts, the historian Mike Davis has estimated, conservatively, that at least 20 million to 30 million people, and probably many more, most of them tropical farmers, perished from the consequences of harsh droughts caused by El Niños and monsoon failures during the nineteenth century, more people than in virtually all the wars of the century.

The Victorian famines are relatively well documented, although, as Davis points out, his fellow historians often ignore them, for the victims who perished were mostly unlettered and their lives unrecorded.[8]

The mortalities for earlier famines, like those that must have occurred during the Medieval Warm Period, are lost to history. An estimated 1.5 million medieval Europeans perished as a result of famine and famine-related diseases during the great rains of 1315–21, which ushered in the Little Ice Age. The casualties among Chinese peasants in the Huang He Valley, among farmers in the coastal river valleys of the Andes, and in the American Southwest must have been significant, especially since many people lived in crowded villages, towns, and pueblos with but rudimentary sanitation. Twentieth-century experience provides a point of comparison. The Chinese famine of 1907 killed an estimated 24 million people. Over 3 million perished in the drought of 1941–42. One and a half million Indians died in the famine of 1965–67, as a result of monsoon failure. Between a quarter of a million and 5 million Russians were victims of a drought in the Ukraine and Volga regions in 1921–22. All of these disasters occurred when global populations were much smaller than today. When one realizes that droughts in the sparsely inhabited Saharan Sahel claimed over 600,000 lives in the droughts of 1972–75 and again in 1984–85, one can only imagine what the magnitudes of these disasters would have been had farming populations been at today's levels.

I hadn't realized until I researched this book just how remarkably flexible human societies were in much of the world a thousand years ago. During the Medieval Warm Period, a high proportion of the earth's population lived in drier environments like those that had nurtured the first civilizations some five thousand years ago. Cities were much smaller, with populations numbering in the tens of thousands, not the millions, which made it easier to adopt strategies for riding out climatic shifts. Every society living in marginal environments for farming developed coping mechanisms. The Ancestral Pueblo of the American Southwest maintained kin ties with distant communities and were prepared to move when droughts came. The Maya stored water at the village and city level. Chimu lords in coastal Peru built elaborate canals and distributed irrigation water with sedulous care. Mande farmers in West Africa's Niger basin developed complex social mechanisms that handled the realities of sudden climatic change, while California Indians stored acorns and depended on their neighbors to make up for food shortfalls.

All of these societies, and others we've described in these pages, were vulnerable to drought. For the most part, with the notable exception of monsoon India and China, they rolled with the climatic punches like trees buffeted by a strong wind. Some environments, like the loess lands of northern China, were so unpredictable that famine was chronic even in years of good rainfall, when floods inundated and saturated thousands of acres of arable land. But in general the human societies of a millennium ago were less vulnerable than we are. There came a point, however, when a society reached a critical mass, a density of urban population and nonfarmers that was unsustainable in drought cycles, or an agricultural economy that had exhausted the land and all opportunities for diversification. The Maya of the southern lowlands of Mexico and Guatemala are a dramatic case in point, where a combination of endemic warfare, rigid governance, environmental degradation, and drought dealt a crippling blow to dozens of towns and great centers. The great lords and the mechanism of state disintegrated and the people dispersed into the subsistence farming villages of a thousand years before. We're dazzled by the splendors of Angkor Wat and Angkor Thom in Cambodia, but, like the Maya, the Khmer lived on the edge of the slippery slope that led downward from self-sustainability.

Fast-forward to the nineteenth century, to the height of Victorian imperial power, where threats to send a gunboat were a powerful diplomatic tool and the technologies of communication and transport—telegraph, steam, rail—infinitely more effective than those of medieval times. A global economy was emerging, founded in great part on grain prices. Yet the nascent vulnerability of the Medieval Warm Period increased a thousandfold. Mike Davis's conservative estimate of twenty million to thirty million casualties sustained, for the most part on the dark, unwritten side of history, beggars the imagination. His figures are from a time when the proportion of the world's population living in semiarid and monsoon lands was a fraction of what it is today.

In the early twenty-first century, some 250 million people live on agriculturally marginal lands, at direct risk from droughts caused by ENSO events, by arbitrary swings of the Southern Oscillation. There are far more instances than those mentioned in earlier chapters. Northeastern Brazil has suffered from major droughts repeatedly, many of them caused by El Niños; Indonesia and Australia are at the mercy of ENSO droughts, to mention only three examples.

Today, the number of people in the world who are highly vulnerable to drought is enormous and growing rapidly, not only in the developing world but also in densely populated areas such as Arizona, California, and southwestern Asia. Judging from the arid cycles of a thousand years ago, the droughts of a warmer future will become more prolonged and harsher. Even without greenhouse gases, the effects of prolonged droughts would be far more catastrophic today than they were even a century ago

Droughts nurture crop failure and allow pastureland to dry out. The same droughts cause rivers to dry up and turn small streams into dry watercourses. Water is the lifeblood of humanity—for agriculture, for herds, for the animals people once hunted, and for drinking. The Chumash Indians of southern California suffered greatly during droughts, not because they lacked food, for they had plenty of fish, but because of a lack of clean water. Crowded into fishing camps and densely inhabited villages near increasingly scarce water supplies, they relied on polluted drinking water, with the inevitable onslaught of disease caused by poor sanitation. The same was true of the British Raj's Indian famine camps in the late nineteenth century, also of ancient eastern Mediterranean cities.

NOW, WITH WARMING accelerating, the stakes for humanity are much higher. Today, we harvest water on an industrial scale—from rainfall, from rivers and lakes, and from rapidly shrinking water tables. Many of us live off looted supplies, brought by aqueduct from the Owens watershed, culled from the Colorado River, and taken from artesian wells, aquifers that will one day run dry. With respect to California, it's sobering to remember that the past seven hundred years were the wettest since the Ice Age. We have experienced droughts, but none of them have endured like those that descended on the Sierra Nevada a millennium ago.

Today, we are experiencing sustained warming of a kind unknown since the Ice Age. And this warming is certain to bring drought—sustained drought and water shortages on a scale that will challenge even small cities, to say nothing of thirsty metropolises like Los Angeles, Phoenix, and Tucson. The Ogallala aquifer, an enormous underground reserve that supplies eight states from Nebraska to Texas, is being depleted at a rate of 42 billion gallons a year. When one hears that an expanding Las Vegas is trying to buy up water supplies from outlying Nevada ranches, one wonders what the future will hold. Will a day come when the hotels on the Strip run out of water because the aquifers have run dry? In terms of water, if the lengthy droughts of a millennium ago were to return, much of the western United States is living on borrowed time.

According to UNESCO, the world has plenty of fresh water, albeit unevenly distributed. But if you look more closely you'll see that, owing to mismanagement, limited resources, and environmental change, almost one fifth of the world's population still lacks access to safe drinking water. Forty percent do not enjoy basic sanitation. In raw figures, UNESCO estimates that 1.1 billion people do not have drinking water supplies, and about 2.6 billion lack basic sanitation.[9] Over half of these people live in China and India, millions more in tropical Africa. These figures come at a time when natural disasters involving water, or its lack, are on the rise. By 2030, UNESCO also estimates, the world will need 55 percent more food, which translates into a growing demand for irrigation, which already claims 70 percent of all fresh water consumed by humans. Then there is the huge increase in urban populations. UNESCO researchers estimate that two thirds of humanity will be urban dwellers by 2030, an estimated 2 billion of them in squatter settlements and slums. The urban poor suffer the most from lack of clean water and sanitation.

The lesson of the Medieval Warm Period for our time is subtle yet alarming. Our journey through the warm and drought-ridden world of a thousand years ago revealed a great diversity of human societies, many of them interconnected by ever-changing economic and sociopolitical ties.

Our travels have taken us down the highways and seaways of a nascent global economy, through a world where interconnectedness and interdependency were beginning to become sustained political realities. We traveled through a time when, on the whole, people lived conservatively, with a good weather eye for risk. Now we confront a future in which most of us live in large and rapidly growing cities, many of them adjacent to rising oceans and waters where Category 5 hurricanes or massive El Niños can cause billions of dollars of damage within a few hours. We're now at a point where there are too many of us to evacuate, where the costs of vulnerability are almost beyond the capacity of even the wealthiest governments to handle. The sheer scale of industrialized societies renders them far more vulnerable to such long-term changes as climbing temperatures and rising sea levels.

This is the immediate crisis of global warming in human terms and it requires not a short-term response but massive intervention on a truly international, and long-term, scale.

We're not good at planning for our great-grandchildren, yet this is what is required of our generation and of those who follow us. There's a political temptation to announce some short-term palliatives and then to claim that we have made a significant contribution to the battle against global warming. Unfortunately, we are past the moment when we can rely on short-term thinking. Drought and water are probably the overwhelmingly important issues for this and future centuries, times when we will have to become accustomed to making altruistic decisions that will benefit not necessarily ourselves but generations yet unborn. This requires political and social thinking of a kind that barely exists today, where instant gratification and the next election seem more important than acting with a view to the long-term future. And a great deal of long-term thinking will have to involve massive investments in the developing world, for those most at risk.

We can't afford to think in provincial terms, of only the drought problems in our own backyards. The warm centuries of a thousand years ago show us that drought is a global problem. Today, we're all interconnected. The experience of the Medieval Warm Period shows how drought can destabilize a society and lead to its collapse. Today, destabilizing forces can jump local boundaries. If we look at how the chance to earn a better living has drawn millions from Latin America across U.S. borders, imagine how many people might uproot themselves if the choice were between famine and food. Many futurists believe the wars of coming centuries will not be fought over petty nationalisms, religion, or democratic principles, but over water, for this most precious of all commodities may become even more valuable than oil. They are probably correct.

How much longer can we remain detached? What will today's casualty figures be like if the droughts projected by the Hadley meteorologists come to pass? They'll be catastrophic, far more so than nineteenth-century fatalities revealed by Mike Davis, and could produce frightening scenarios. Are we looking, for example, at a time when enormous, uncontrollable mass migrations of people fleeing hunger and drought will burst across territorial boundaries? Such population movements are not beyond the realm of possibility.

It's been easy for us to forget that millions of people still live at the subsistence level and use basically medieval technologies to wrest a living from the soil. We can no longer afford benign ignorance, for the long-term perils of chronic drought connect all humankind in ways that we are only just beginning to understand. In an earlier book, I described industrial society as a huge supertanker that takes many miles to stop and maneuvers slowly.[10] I accused our society of being oblivious and inattentive, of ignoring the climatic danger signals that lie ahead.

Thanks to a new generation of science and thanks to activists ranging from Al Gore to university students, global warming has become a political issue and a topic of fascination for the chatterers. Yet, it's striking, and very frightening, that the elephant of drought is still so widely ignored.

History is always around us, threatening, offering encouragement, sometimes showing us precedents. The warm centuries of a thousand years ago remind us that we have never been masters of the natural world; at our best, we have accommodated ourselves to its fickle realities. As the Khmer and the Maya show us, the harder we try to master it, the greater our risk of sliding down the hazardous slope of unsustainability. We should accept this reality and not be frightened by a future where we are not the masters; we must cease trying to assume that role. The people of a thousand years ago remind us that our greatest asset is our opportunism and endless capacity to adapt to new circumstances. Let us think of ourselves as partners with rather than potential masters of the changing natural world around us.

Acknowledgments

The Great Warming is the culmination of over a decade of thinking and writing about ancient climate change. It treats of a subject that is still little known, information about which is scattered in obscure academic literatures in dozens of languages. I've attempted to produce a synthesis that is very much my own take on a confusing jigsaw puzzle of archaeology, history, and paleoclimatology. I am, of course, responsible for the conclusions and accuracy of this book, and, no doubt, will hear in short order from those kind, often anonymous individuals who delight in pointing out errors large and small. Let me thank them in advance.

The research for this book involved consulting a large number of busy specialists, who were invariably courteous and often startlingly prompt in their replies. I am deeply grateful to them for taking my importunings seriously. It's impossible to name everyone, but my long list of indebtedness includes Reid Bryson, Roseanne D'Arrigo, Carole Crumley, Ronald Fletcher, Michael Glantz, Michael Glassow, John Johnson, Doug Kennett, Ian Lindsay, Roderick McIntosh, George Michaels, Dan Penny, Mark Rose, Vernon Scarborough, Chris Scarre, Scott Stine, and Stan Wolpert.

In Santa Barbara, Shelly Lowenkopf was my passionate foil and the best of writing coaches. He saw me through many tricky moments and was, as always, enthusiastic and supportive. Steve Brown drew the maps and drawings with his customary skill. Bill Frucht was extremely helpful in the early stages of the book. My debt to him is enormous. My agent, Susan Rabiner, was a tower of strength and inspiration.

Peter Ginna and Katie Henderson provided editorial guidance when the manuscript was at its most critical stages. In a real sense, this is their book as much as mine. Their merciless and perceptive criticism made the final stages of writing much easier. I'm deeply grateful.

Finally, and as always, my thanks to Lesley, Ana, and our menagerie, who have suffered through the pains of writing with amused tolerance. In fact, I've just been delayed for five minutes by a cat sitting on the keyboard demanding attention. Life is always interesting around here.

Brian Fagan
Santa Barbara, California

Notes

The literature on the societies described in these pages is enormous and growing rapidly, but is dwarfed by the current tidal wave of publications on ancient and modern climate change. New journals and books appear daily. Many of them are, of course, highly specialized and of little relevance to these pages. The references that appear below provide a good cross-section of the literature as of mid-2007 and contain useful bibliographies for those wishing to probe deeper.

Preface

1. For the benefit of those confused by the terms "climate" and "weather," climate is the accumulation of daily and seasonal weather events over a long period of time. Weather is the state of the atmosphere in terms of such variables as temperature, cloudiness, rainfall, and radiation at a moment in time. In other words, climate is cumulative experience, weather is what you get.

2. Lamb's "Medieval Warm Period" is the most commonly used term to describe what he perceived as a period of medieval warming. Many climatologists rightly question whether the term has global validity, on the grounds that it is not well defined and was, in fact, a period of highly variable climatic conditions. Others use the term "Medieval Climatic Anomaly." In the interest of clarity, I have used "Medieval Warm Period" to the exclusion of other terms, although I occasionally refer to "the warm centuries" as a generic term, even if the centuries weren't all warm. Experts may cavil, but, after all, "Medieval Warm Period" is convenient, in common use, and widely known. And most serious students of the subject know that the term is something of a misnomer.

3. Brian Fagan, *The Little Ice Age* (New York: Basic Books, 2000).

4. Mike Davis, *Late Victorian Holocausts* (New York: Verso, 2001). Preface.

Chapter 1: A Time of Warming

1. Hubert Lamb, *Climate History and the Modern World* (London: Methuen, 1982), p. 173.

2. Discussion in Bruce C. Campbell, "Economic Rent and the Intensification of English Agriculture, 1086–1350." In Grenville Astill and John Langdon, eds., *Medieval Farming and Technology* (Leiden: Brill, 1997), pp. 225–50. The essays in this book are a mine of information on medieval agriculture, its yields, and changing technology.

3. For the Mount Tambora disaster, see Henry Stommel and Elizabeth Stommel, *Volcano Weather: The Story of 1816, the Year Without a Summer* (Newport, R.I.: Seven Seas Press, 1983).

4. George S. Philander, *Is the Temperature Rising?* (Princeton: Princeton University Press, 2000), p. 125.

5. William of Malmesbury (c. 1096–1143) was a monk at Malmesbury in southwestern England and a medieval historian rated second only to the Venerable Bede. His greatest works were *Gesta regum Anglorum*, a history of the English kings from 449 to 1127, and *Historia novella*, which continues the story. Book V covers contemporary history, which is where his vineyard observations are to be found.

6. Lamb, *Climate History*, chapter 10.

7. Hubert Lamb and Knud Frydendahl, *Historic Storms of the North Sea, British Isles and Northwestern Europe* (Cambridge, Eng.: Cambridge University Press, 1991).

8. Emmanuel Le Roy Ladurie, *Times of Feast, Times of Famine: A History of Climate Since the Year 1000*. Barbara Bray, trans. (Garden City, N.Y.: Doubleday, 1971).

9. Lamb, *Climate History*, chapter 10.

10. Hereward the Wake (fl. 1070) is one of the great, but little-known heroes of early English history. We know he was exiled by the Saxon king Edward the Confessor in 1062, returning after 1066 to find his father dead, his brother murdered, and a Norman lord in possession of his home. He became a rallying point for resistance against William the Conqueror. Hereward attacked and sacked Peterborough Abbey in 1070 with the help of a Danish army. When William bribed the Danes to return home, Hereward continued his revolt from a stronghold at Ely. When William finally captured his redoubt, he fled into hiding. His ultimate fate is unknown.

11. Lamb and Frydendahl, *Historic Storms*, p. 34, and Brian Fagan, *Fish on*

Friday: Feasting, Fasting, and the Discovery of the New World (New York: Basic Books, 2006), p. 101.

12. M. E. Mann, R. S. Bradley, and M. K. Hughes, "Global Surface Temperature Patterns and Climate Forcing over the Past 6 Centuries," *Nature* 392 (1998), pp. 779–87, and M. E. Mann, R. S. Bradley, and M. K. Hughes, "Northern Hemisphere Temperatures During the Past Millennium: Inferences, Uncertainties, and Limitations," *Geophysical Research Letters* vol. 26, no. 6 (1999), pp. 759–62.

13. Intergovernmental Panel on Climate Change, *Climate Change 2001: The Scientific Basis* (Cambridge, Eng.: Cambridge University Press, 2001).

14. For a discussion, see National Research Council, *Surface Temperature Reconstructions for the Past 2,000 Years* (Washington, D.C.: National Academies Press, 2006), pp. 1ff.

15. Information from Lamb, *Climate History*, pp. 169–70.

Chapter 2: "The Mantle of the Poor"

1. Tertullian, *De Anima*, vol. 1, XXX.210.

2. For a general description of the period, see William Jordan, *Europe in the High Middle Ages* (New York: Viking, 2001). I have drawn extensively on this work here.

3. Kenneth Clark, *Civilization: A Personal View* (New York: Harper & Row, 1969), p. 23.

4. Population: Jordan, *Europe*, pp. 7–10.

5. Bruce C. Campbell, "Economic Rent and the Intensification of English Agriculture, 1086–1350." In Grenville Astill and John Langdon, eds., *Medieval Farming and Technology* (Leiden: Brill, 1997), pp. 225.

6. Medieval agricultural technology is a complex subject and generalization is difficult. A summary: Georges Comet, "Technology and Agricultural Expansion in the Middle Ages: The Example of France North of the Loire." In Astill and Langdon, *Medieval Farming*, pp. 11–40.

7. This paragraph is based on Campbell, "Economic Rent," pp. 233ff.

8. Paragraph based on Jordan, *Europe*, pp. 16–17; quote from p. 17. For Southwark, see Martha Carlin, *Medieval Southwark* (Rio Grande, Oh.: Hambledon, 1996), pp. 250–51.

9. Michael Williams, *Deforesting the Earth: From Prehistory to Global Crisis* (Chicago: University of Chicago Press, 2003), is the definitive work on the subject. I drew on it extensively here for medieval deforestation.

10. Land reclamation: Grenville Astill, "Agricultural Production and Technology in the Netherlands, c. 1000–1500." In Astill and Langdon, *Medieval Farming*, pp. 89–114.

11. Williams, *Deforesting*, p. 105.

12. This hypothetical scenario is based on Williams, *Deforesting*, chapter 5 and my own central African forest-clearance experience with subsistence farmers.

13. Pagans: quoted from Williams, *Deforesting*, p. 122.

14. Williams, *Deforesting*, p. 111.

15. Jean Leclercq, *The Love of Learning and the Desire for God: Study of Monastic Culture*. Catherine Misrahi, trans. (London: S.P.C.K., 1978). p. 60.

16. Quoted from James Westfall Thompson, *An Economic and Social History of the Middle Ages, 300 to 1300* (New York: Century, 1908), p. 611.

17. This passage is based on Fagan, *Fish on Friday: Feasting, Fasting, and the Discovery of the New World* (New York: Basic Books, 2006), chapters 3, 4 and 7, where primary references will be found.

18. Fagan, *Fish on Friday*, chapters 5 and 6.

19. William Chester Jordan, *The Great Famine* (Princeton: Princeton University Press, 1996), is a superb account of this catastrophic event.

20. Quote from Martin Bouquet et al., eds., *Receuil des Historiens des Gaules et de la France* (1738–1904) 21:197.

21. A recent account of the Black Death: John Kelly, *The Great Mortality* (New York: HarperCollins, 2005).

Chapter 3: The Flail of God

1. Ch'ang Ch'un, *The Travels of an Alchemist Recorded by His Disciple Li Chih-Ch'ang*. Arthur Waley, trans. (London: Routledge, 1931), p. 104.

2. John Trevisa's translation of Bartholomew the Englishman's *Latin Encyclopaedia*. See R. Barber, *The Penguin Guide to Medieval Europe* (London: Penguin Books, 1984), p. 30. The conventional boundary between Europe and Asia is now defined as the Ural Mountains.

3. Ginghis Khan, sometimes called Chingis Khan or Chingiss Khan, and commonly known as Genghis Khan, has been the subject of numerous books. Leo de Hartog, *Khan: Conqueror of the World* (London: I. B. Tauris, 1989), is a lucid account. See also George Lane, *Genghis Khan and Mongol Rule* (Westport, Conn.: Greenwood Press, 2004).

4. J. A. Boyle, trans., *Tarikh-I Jahan Gusha*. In *The History of the World Conqueror* (Manchester: Manchester University Press, 1997), p. 105.

5. Extracts from *The Chronicle of Novgorod* may be found at http://www.fordham.edu/halsall/source/novgorod1.html. The quote is from the entry for A.D. 1238.

6. Boyle, *History*, p. 105.

7. 'Ata Malik Jowanyi (Juvaini) (1226–1283) served at the Mongol court from childhood. He became a historian and later governor of Baghdad in about 1260. His history has considerable credibility as he witnessed many of the events described therein.

8. Maria Shahgedanova, ed., *The Physical Geography of Northern Eurasia* (New York: Oxford University Press, 2002), offers a series of technical summaries of central Asian environments. For the steppes, I relied on Alexander Chibilyov, "Steppe and Forest-Steppe," pp. 248–66.

9. Willem van Ruysbroeck, *The Mission of Friar William of Rubreck: His Journey to the Court of the Great Khan Möngke, 1253–1255* (London: Hakluyt Society, 1990), p. 118.

10. Research summarized in E. M. Lavenko and Z. V. Karamysheva, "Steppes of the Former Soviet Union and Mongolia." In R. T. Coupland, ed., *Natural Grasslands: Eastern Hemisphere and Résumé. Ecosystems of the World*, vol. 8b (London: Elsevier, 1979), pp. 3–60.

11. An excellent summary of the domestication of the horse: David W. Anthony, "The 'Kurgan Culture': Indo-European Origins, and the Domestication of the Horse: A Reconsideration," *Current Anthropology*, vol. 27, no. 4 (1986), pp. 291–313.

12. A useful, albeit somewhat outdated, account of the Scythians: Tamara Rice, *The Scythians* (London: Thames & Hudson, 1957).

13. Herodotus, *The Histories*. Robin Waterfield, trans. (Oxford, Eng.: Oxford University Press, 1998), 4:127, p. 277.

14. This section draws on Ian Blanchard, "Cultural and Economic Activities in the Nomadic Societies of the Trans-Pontine Steppe," *Annual of Medieval Studies at CEU*, vol. 11 (2005).

15. An excellent summary, with superb photographs, of modern Mongolian nomad life at a time of major political change can be found in Melvyn C. Goldstein and Cynthia M. Beall, *The Changing World of Mongolia's Nomads* (Berkeley: University of California Press, 1994).

16. Elena E. Kuz'mina, "Stages of Development of Stock-Breeding Husbandry and Ecology of the Steppes in the Light of the Archaeological and Palaeoecological Data (4th Millennium BC–8th Century BC)." In Bruno Genito, ed., *The Archaeology of the Steppes: Methods and Strategies* (Naples: Instituto Universitario Orientale, 1994), vol. 44, pp. 31–72.

17. *Chronicle of Novgorod*, 1230; see http://www.fordham.edu/halsall/source/novgorod1.html.

18. Rosanne D'Arrigo et al., "1738 Years of Mongolian Temperature Variability Inferred from a Tree-Ring Width Chronology of Siberian Pine," *Geophysical Research Letters*, vol. 28, no. 2 (2001), pp. 543–46.

19. Lane, *Genghis Khan*, p. 45.

20. Robert Cowley, ed., *What If?* (New York: Berkley Trade, 2000).

Chapter 4: The Golden Trade of the Moors

1. Anonymous, *Toffut-al-Alabi* (12th century). Quoted from H. R. Palmer, *Sudanese Memoirs: Being Mainly Translations of a Number of Arabic Manuscripts Relating to the Central and Western Sudan* (Lagos, Nigeria: Government Printer, 1928), vol. 2, p. 90.

2. Roderick J. McIntosh, "Chasing Dunjugu over the Mande Landscape: Making Sense of Prehistoric and Historic Climate Change," *Mande Studies*, vol. 6 (2004), pp. 11–28. I have drawn heavily on this important paper for the climatic scenario presented here. See also: Robin Dunbar, "Climate Variability During the Holocene: An Update." In Roderick J. McIntosh, Joseph A. Tainter, and Susan Keech McIntosh, eds., *The Way the Wind Blows: Climate, History, and Human Action* (New York: Columbia University Press, 2000), pp. 45–88.

3. Sharon E. Nicholson, "Recent Rainfall Fluctuations in Africa and Their Relationship to Past Conditions Over the Continent," *The Holocene*, vol. 4 (1994), pp. 121–31. See also S. E. Nicholson and J. P. Grist, "A Conceptual Model for Understanding Rainfall Variability in the West African Sahel on Interannual and Interdecadal Timescales," *International Journal of Climatology*, vol. 21 (2001), pp. 1733–57.

4. Gerald Haug et al., "Southward Migration of the Intertropical Convergence Zone through the Holocene," *Science*, vol. 293 (2001), pp. 1304–7.

5. Herodotus, *The Histories*. Robin Waterfield, trans. (Oxford: Oxford University Press, 1998), 3:32.

6. Endless academic controversy surrounds putative Roman visits to West Africa, but such contacts, if any, were fleeting at best. For a much respected summary of the Saharan trade generally, see E. W. Bovill, *The Golden Trade of the Moors*, rev. ed. (London: Oxford University Press, 1968), p. 121.

7. Camels were first domesticated in Arabia in about 1500 B.C., but did not come into common use until the centuries before Christ. Everything depended on

the saddle. At first, camel riders sat on a saddle mounted on the animal's hindquarters. They used sticks to control their beasts and were so close to the ground that they lost a major advantage of the camel: its height. The north Arabian saddle, a rigid structure mounted over the hump, changed the equation during the five centuries before Christ. Its owner could now carry a modest load and fight from the saddle with sword or spear. So effective was the north Arabian saddle that wheeled carts effectively vanished from southwestern Asia for many centuries. A classic study: Richard W. Bulliet, *The Camel and the Wheel* (Cambridge, Mass.: Harvard University Press, 1975).

8. Quoted from Ian Blanchard, *Mining, Metallurgy and Minting in the Middle Ages*, vol. 1 (Stuttgart: Franz Steiner, 2001), p. 156.

9. Extended discussion in ibid., pp. 91–102.

10. Ibid., pp. 153–54.

11. Nehemiah Levetzion, *Ancient Ghana and Mali* (London: Methuen, 1973), p. 189.

12. According to Bovill, *Golden Trade*, p. 81, a nugget twice this size came from Bambuk in about 1900.

13. Roderick J. McIntosh, *The Peoples of the Middle Niger: The Island of Gold* (Oxford, Eng.: Blackwell, 1988), pp. 257–59.

14. Ibid., pp. 267–81.

15. Ibid., p. xv.

16. This section draws heavily on Roderick J. McIntosh, "Social Memory in Mande." In McIntosh et al., *The Way the Wind Blows*, pp. 141–80. I also consulted McIntosh, *Peoples of the Middle Niger*, and Téréba Togola, "Memories, Abstraction, and Conceptualization of Ecological Crisis in the Mande World." In McIntosh et al., *The Way the Wind Blows*, pp. 181–92. All three references contain excellent specialist bibliographies

17. Susan K. McIntosh, *Prehistoric Investigations in the Region of Jenne, Mali*. 2 vols. (Oxford, Eng.: British Archaeological Reports, 1980). See also the same author's "Results of Recent Excavations at Jenné-jeno and Djenné, Mali," *Proceedings of the 11th Panafrican Congress of Prehistory and Related Studies, Bamako* (Bamako: Institut des Sciences Humaines, 2005), pp. 115–22.

18. R. McIntosh, "Social Memory in Mande." In McIntosh et al., *The Way the Wind Blows*, pp. 141–80.

19. The Almoravids were members of the al-Murabitum cult of Islam, who waged jihad in the western Sahara in the eleventh century under Ibn Yasin and later Abu Bakr, who captured Koumbi. See Nehemiah Levetzion, *Ancient Ghana and Mali* (London: Methuen, 1973); also, Bovill, *Golden Trade*, chapter 7.

Chapter 5: Inuit and Qadlunaat

1. Magnus Magnusson and Herman Palsson, eds., *The Vinland Sagas: The Norse Discovery of America* (London: Penguin Books, 1965), p. 55.

2. Common usage is to call Canadian Arctic peoples Inuit, those in Alaska and the Bering Strait region Eskimo. I have followed this convention here.

3. More on Eskimo and Inuit. The term "Eskimo," used by the peoples of Alaska and Siberia, is said to be derived from an Algonkin Indian expression: "People Who Eat Raw Meat." (Or perhaps "The People Who Live Up the Coast.") Some people consider the word derogatory, so Canadians usually use "Inuit," which simply means "Humans" in local speech. For the purposes of this book, I have followed Robert McGhee's pragmatic usage, using "Inuit" for all Eskimos living in northern Alaska, Arctic Canada, and Greenland. They share a language, Inuktituut. Robert McGhee, *The Last Imaginary Place: A Human History of the Arctic World* (Oxford, Eng.: Oxford University Press, 2006), p. 104.

4. The best summary of Norse voyaging for the general reader is William W. Fitzhugh and Elisabeth I. Ward, eds., *Vikings: The North Atlantic Saga* (Washington, D.C.: Smithsonian Institution Press, 2000), in which authoritative essays on all aspects of the subject will be found, as will an excellent bibliography.

5. Kirsten Seaver, *The Frozen Echo* (Stanford: Stanford University Press, 1996), is a widely quoted source on Norse Greenland.

6. A summary of this site will be found in Brian Fagan, *Ancient North America*, 4th ed. (London: Thames & Hudson, 2004), chapter 1.

7. Magnusson and Palsson, *Vinland Sagas*, p. 99.

8. The hunting cultures of the Bering Strait are summarized in my *Ancient North America*, chapters 8 and 9, which should be read in conjunction with McGhee's *Last Imaginary Place.* A highly technical and admirably complete analysis will be found in Owen K. Mason, "The Contest Between the Ipiutak, Old Bering Sea, and Birnik Polities and the Origin of Whaling During the First Millennium A.D. Along Bering Strait," *Journal of Anthropological Archaeology*, vol. 17 (1998), pp. 240–325.

9. See the discussion in Mason, "Contest," p. 250ff.

10. Helge Larsen and E. Rainey, "Ipiutak and the Arctic Whale Hunting Culture," *Anthropological Papers of the American Museum of Natural History*, vol. 42 (1948).

11. On Ekven, see McGhee, *The Last Imaginary Place*, pp. 218–19.

12. Mikhail Bronshtein and Patrick Plumet, "Ékven: L'Art Préhistorique Béringien et l'Approache Russe de l'Origine de la Tradition Culturelle Esquimaude," *Études/Inuit/Studies*, vol. 19, no. 2 (1995), pp. 5–59.

13. McGhee, *The Last Imaginary Place*, p. 119ff.

14. Robert McGhee, *Ancient People of the Arctic* (Vancouver: University of British Columbia Press, 1996), has an excellent description of Tuniit culture.

15. Moreau Maxwell, *The Prehistory of the Eastern Arctic* (New York: Academic Press, 1985), p. 222.

16. Heather Pringle, "New Respect for Metal's Role in Ancient Arctic Cultures," *Science*, vol. 277 (1997), pp. 766–67.

17. Therkel Mathiassen, "Archaeology of the Central Eskimos, the Thule Culture and Its Position Within the Eskimo Culture," *Report of the Fifth Thule Expedition, 1921–1924* (Copenhagen: Glygenclalski Boghandel, Nordisk Forlang, 1927).

18. Discussion in McGhee, *The Last Imaginary Place*, p. 121ff.

19. Robert McGhee, "Contact Between Native North Americans and the Medieval Norse: A Review of Evidence," *American Antiquity*, vol. 49 (1984), pp. 4–26.

20. On the Skraelings, see McGhee, *The Last Imaginary Place*, p. 93.

21. This section is based on Seaver, *The Frozen Echo*.

22. Brian Fagan, *The Little Ice Age* (New York: Basic Books, 2000), chapter 1.

Chapter 6: The Megadrought Epoch

1. Malcolm Margolin, ed., *The Way We Lived* (Berkeley: California Historical Society and Heyday Books, 1993), p. 125.

2. Scott Stine, "Extreme and Persistent Drought in California and Patagonia During Mediaeval Time," *Nature*, vol. 369 (1994), pp. 546–49.

3. Celine Herweijer et al., "North American Droughts of the Last Millennium from a Gridded Network of Tree-Ring Data," *Journal of Climate*, vol. 20, no. 7 (2007), pp. 1353–76. This important paper summarizes the evidence for drought over a wide area. I drew heavily on it here.

4. Ibid.

5. A sidebar on ENSO, El Niños, and La Niñas appears in chapter 9.

6. A summary of Great Basin archaeology will be found in Brian Fagan, *Ancient North America*, 4th ed. (London: Thames & Hudson, 2004), chapter 12. See also Donald L. Grayson, *A Natural History of the Great Basin* (Washington, D.C.: Smithsonian Institution Press, 1993). For the impact of Medieval Warm Period drought, see Terry L. Jones et al., "Environmental Imperatives Reconsidered: Demographic Crises in Western North America During the Medieval Climatic Anomaly," *Current Anthropology*, vol. 40, no. 2 (1999), pp. 137–70.

7. C. Melville Aikens, *Hogup Cave. University of Utah Anthropologial Papers*, vol. 92, (1970).

8. Summarized in Kurt Repamshek, "Shelter from the Prehistoric Storm," *American Archaeology*, vol. 11, no. 1 (2007), pp. 26–32.

9. For Great Basin native American groups, see William C. Sturtevant, ed., *Handbook of North American Indians.* Vol. 11: *Great Basin*, ed. Warren L. d'Azevedo (Washington, D.C.: Smithsonian Institution, 1986).

10. See http://danr.ucop.edu/ihrmp/oak27.htm.

11. See Lowell J. Bean, *Mukat's People: The Cahuilla Indians of Southern California* (Berkeley: University of California Press, 1972).

Chapter 7: Acorns and Pueblos

1. Alfred L. Kroeber, *Handbook of the Indians of California* (Washington, D.C.: Bureau of American Ethnology, 1925), p. 524.

2. Travis Hudson, Jan Timbrook, and M. Rempe, *Tomol: Chumash Watercraft as Described in the Ethnographic Notes of John P. Harrington* (Los Altos and Santa Barbara, Calif.: Ballena Press and Santa Barbara Museum of Natural History, 1978), p. 22.

3. A useful summary of the impact of acorns: Mark A. Basgall, "Resource Intensification Among Hunter-Gatherers: Acorn Economies in Prehistoric California," *Research in Economic Anthropology*, vol. 9 (1987), pp. 21–52.

4. Sarah Mason, "Acorntopia? Determining the Role of Acorns in Past Human Subsistence." In John Wilkins, David Harvey, and Michael Dobson, eds., *Food in Antiquity* (Exeter, U.K.: University of Exeter Press, 1995), pp. 112–36.

5. Beverly R. Ortiz, as told to Julia Parker, *It Will Live Forever: Traditional Yosemite Acorn Preparation* (Berkeley: Heyday Books, 1991). See also, Walter Goldschmidt, "Nomlaki Ethnography," *University of California Publications in American Archaeology and Ethnology*, vol. 42, no. 4 (1951): 303–443.

6. Pat Mikkelsen, William Hildebrandt, and Deborah Jones, "Toolstone Procurement and Lithic Production Technology, California." In Michael Moratto, ed., *Archaeological Investigations PGT–PG&E Pipeline Expansion Project, Idaho, Washington, and California.* Vol. 4 (Report Submitted to the Pacific Gas Transmission Company, Portland, Oregon, 1994), chapter 8. For California trade generally, see Brian Fagan, *Before California* (Lanham, Md.: Rowman and Littlefield, 2003), chapter 7.

7. See http://danr.ucop.edu/ihrmp/oak27.htm.

8. Elizabeth Weiss, "Drought-Related Changes in Two Hunter-Gatherer California Populations," *Quaternary Research*, vol. 58 (2002), pp. 393–96.

9. Section based on Douglas J. Kennett, *Behavioral Ecology and the Evolution of Hunter-Gatherer Societies on the Northern Channel Islands, California* (Berkeley: University of California Press, 1998).

10. Douglas J. Kennett and James P. Kennett, "Competitive and Cooperative Responses to Climatic Instability in Coastal Southern California," *American Antiquity*, vol. 65 (2000), pp. 379–95.

11. Summary in Fagan, *Before California*, chapter 14.

12. Jeanne Arnold, ed., *Origins of a Pacific Coast Chiefdom* (Berkeley: University of California Press, 2001), is a fundamental source on Santa Cruz Island.

13. Patricia M. Lambert and Phillip L. Walker, "Physical Anthropological Evidence for the Evolution of Social Complexity in Coastal Southern California," *Antiquity*, vol. 65 (1991), pp. 963–73.

14. Drills described by Arnold, *Origins*.

15. Tessie Naranjo, "Thoughts on Migration by Santa Clara Pueblo," *Journal of Anthropological Archaeology*, vol. 14 (1995), pp. 247–50.

16. An enormous literature surrounds Chaco Canyon. For a popular account, see Brian Fagan, *Chaco Canyon: Archaeologists Explore the Lives of an Ancient Society* (New York: Oxford University Press, 2005). See also Steve Lekson, ed., *The Archaeology of Chaco Canyon: An Eleventh-Century Pueblo Regional Center* (Santa Fe: School of American Research, 2006).

17. Gwinn Vivian, *Chacoan Prehistory of the San Juan Basin* (San Diego: Academic Press, 1990), p. 432ff.

18. Gwinn Vivian, "Chaco Roads: Morphology," *Kiva*, vol. 63, no. 1 (1997), pp. 7–34, and "Chaco Roads: Function," *Kiva*, vol. 63, no. 1 (1997), pp. 35–67.

19. This section is based on Mark D. Varien and Richard H. Wilshusen, eds., *Seeking the Center Place: Archaeology and Ancient Communities in the Mesa Verde Region* (Salt Lake City: University of Utah Press, 2002).

Chapter 8: Lords of the Water Mountains

1. Dennis Tedlock, trans., *Popol Vuh: The Mayan Book of the Dawn of Life* (New York: Touchstone, 1996), p. 71. The *Popol Vuh*, or "Council Book," is a book of creation originally written in Maya glyphs, and one of the masterpieces of native American literature. The Quiché Maya live in the Guatemalan highlands.

2. Gerald H. Haug et al., "Climate and the Collapse of Maya Civilization," *Science*, vol. 299 (2003): 1731–35.

3. David A. Hodell et al., "Possible Role of Climate in the Collapse of Classic Maya Civilization," *Nature*, vol. 375 (1995), pp. 391–94.

4. Tedlock, *Popol Vuh*, p. 64.

5. Accounts of Maya civilization abound. The most widely read: Michael D. Coe, *The Maya*, 7th ed. (London: Thames & Hudson, 2005).

6. This section is based on Vernon L. Scarborough, "Ecology and Ritual: Water Management and the Maya," *Latin American Antiquity*, vol. 9, no. 2 (1998), pp. 135–59. See also the same author's more general work *The Flow of Power: Ancient Water Systems and Landscapes* (Santa Fe: SAR Press, 2003).

7. For El Mirador, see Ray Matheny, *El Mirador, Petén: An Interim Report* (Provo, Ut.: New World Archaeological Foundation Papers, 1980).

8. R. T. Matheny et al., *Investigations at Edzná, Campeche, Mexico*. Vol 1, Part 1: *The Hydraulic System* (Provo, Ut.: New World Archaeological Foundation Papers, 1983).

9. For Tikal, see P. D. Harrison, *The Lords of Tikal: Rulers of an Ancient Maya City* (New York: Thames & Hudson, 1999). A popular, but sometimes controversial, account of Maya civilization based on archaeology and glyphs: Linda Schele and David Freidel, *A Forest of Kings* (New York: William Morrow, 1990).

10. The estimate of 210 acres (85 hectares) is based on calculations from a study of modern irrigation at Chilac, near Tehuacán. G. C. Wilken, *Good Farmers* (Berkeley: University of California Press, 1987).

11. Scenario based on Schele and Freidel, *Forest*, pp. 280–81.

12. Patricia McAnany, *Living with the Ancestors: Kinship and Kingship in Ancient Maya Society* (Austin: University of Texas Press, 1995).

13. A huge literature surrounds the collapse of ancient Maya civilization. An excellent summary will be found in David Webster, *The Fall of the Ancient Maya* (London: Thames & Hudson, 2002).

14. Schele and Freidel, *Forest*, chapter 8.

15. An excellent summary of Copán and its end will be found in David L. Webster et al., *Copán: The Rise and Fall of an Ancient Maya Kingdom* (New York: Wadsworth, 1999), and E. Wyllis Andrews IV and William L. Fash, *Copán* (New York: James Currey, 2005).

16. The Petexbatun project and Maya civilization generally: Arthur Demarest, *The Rise and Fall of a Rainforest Civilization* (New York: Cambridge University Press, 2005).

Chapter 9: The Lords of Chimor

1. Father Bernabé Cobo (1580–1657) was a Jesuit missionary and scholar who spent sixty-one years in Peru. His main work was *Historia general de las In-*

dias, completed in 1653. Only the first half survives. Cobo was a perceptive observer. His work is a primary source on early Peru and its peoples. Quote from his *Inca Religion and Customs*, Roland Hamilton, trans. (Austin: University of Texas Press, 1990), p. 123. Cobo's observations on Inca royal mummies can equally be applied to those of the lords of Chimor.

2. Oca, *Oxalis tuberosa*, is a root vegetable with a starchy, edible tuber, widely grown in the Andes in a wide variety of colors. It grows in the same environments as ullucu (*Ullucus tuberosus*), a cool-climate tuber. Both were frozen or dried and widely traded.

3. Ephraim Squier, *Travels in Peru* (New York: Harpers, 1888), p. 110.

4. L.Thompson et al., "A 1,500-Year Tropical Ice Core Record of Climate: Potential Relations to Man in the Andes," *Science*, vol. 234 (1986), pp. 361–64.

5. Alan Kolata, "Environmental Thresholds and the 'Natural History' of an Andean Civilization." In Garth Bawdon and Richard Martin Reycraft, eds., *Environmental Disaster and the Archaeology of Human Response* (Albuquerque: Maxwell Museum of Anthropology, 2000), pp. 195–212.

6. A summary of the Tiwanaku drought will also be found in Brian Fagan, *The Long Summer* (New York: Basic Books, 2004), chapter 12.

7. Scott Stine, "Extreme and Persistent Drought in California and Patagonia in Mediaeval Time," *Nature*, vol. 369 (1994), pp. 546–49.

8. Scott Stine and M. Stine, "A Record from Lake Cardiel of Climate Change in Southern South America," *Nature*, vol. 345 (1990), pp. 705–8.

9. Michael Glantz, *Currents of Change* (New York: Cambridge University Press, 1996), offers a widely read account of El Niño. See also Brian Fagan, *Floods, Famines and Emperors: El Niño and the Collapse of Civilizations* (New York: Basic Books, 1999).

10. The literature is summarized in Mary Van Buren, "The Archaeology of El Niño Events and Other 'Natural' Disasters," *Journal of Archaeological Method and Theory*, vol. 8, no. 2 (2001), pp. 129–49.

11. B. Rein et al., "A Major Holocene ENSO Anomaly During the Medieval Period," *Geophysical Research Letters*, vol. 31, no. 10 (2004), p. L17211. On Ecuador, see Christopher M. May et al., "Variability of El Niño: Southern Oscillation Activity at Millennial Timescales During the Holocene Epoch," *Nature*, vol. 470 (2002), pp. 162–65.

12. A hypothetical scenario, but the details reflect the spectacular regalia of two Moche lords buried at Sipán in the lower Lambayeque Valley in about A.D. 400. Walter Alva and Christopher Donnan, *The Royal Tombs of Sipán* (Los Angeles: UCLA Fowler Museum of Cultural History, 1993).

13. The best general summary of Andean archaeology and history is Michael

Moseley, *The Incas and Their Ancestors*, 2nd ed. (New York: Thames & Hudson, 2000).

14. Michael Moseley and Kent C. Day, eds., *Chan Chan: Andean Desert City* (Albuquerque: University of New Mexico Press, 1982), is the authoritative report.

15. Figure from Moseley, *Incas*, p. 256.

16. This section is based on Van Buren, "Archaeology," and more specialist references.

Chapter 10: Bucking the Trades

1. T. Gladwin, *East Is a Big Bird: Navigation and Logic on Puluwat Atoll* (Cambridge, Mass.: Harvard University Press, 1972), p. 87.

2. An account of the collapse of Rapa Nui can be found in Jared Diamond, *Collapse* (New York: Viking, 2005), chapter 2. A summary of Easter Island archaeology: Paul Bahn and John Flenley, *Easter Island, Earth Island* (London: Thames & Hudson, 1992).

3. Ben Finney, *Voyage of Rediscovery: A Cultural Odyssey through Polynesia* (Berkeley: University of California Press, 1994), p. 3.

4. Edwin Clark et al., "Evidence for a 'Medieval Warm Period' in a 1,100-Year Tree-Ring Reconstruction of Past Austral Summer Temperatures in New Zealand," *Geophysical Research Letters*, vol. 29, no. 14 (2002), pp. 12(1)–12(4).

5. Patrick D. Nunn, "Environmental Catastrophe in the Pacific Islands Around A.D. 1300," *Geoarchaeology*, vol. 16, no. 7 (2000), pp. 715–40.

6. Kim M. Cobb et al., "El Niño/Southern Oscillation and Tropical Pacific Climate During the Last Millennium," *Nature*, vol. 724 (2003), pp. 271–75. Also see Kim M. Cobb, Christopher D. Charles, and David E. Hunter, "A Central Tropical Pacific Coral Demonstrates Pacific, Indian, and Atlantic Decadal Connections," *Geophysical Research Letters*, vol. 18, no. 11 (2001), pp. 2209–12.

7. See B. Rein et al., "A Major Holocene ENSO Anomaly During the Medieval Period," *Geophysical Research Letters*, vol. 31, no. 10 (2004), p. L172211, and Christopher M. May et al., "Variability of El Niño: Southern Oscillation Activity at Millennial Timescales During the Holocene Epoch," *Nature*, vol. 470 (2002), pp. 162–165.

8. James Beaglehole, *Captain James Cook: A Life* (Stanford: Stanford University Press, 1974), p. 178.

9. James Beaglehole, ed., *The Journals of Captain James Cook on His Voyages of Discovery* (London: Hakluyt Society, 1968), p. 354.

10. Patrick Vinton Kirch, *On the Road of the Winds* (Berkeley: University of

California Press, 2000), describes Lapita and the archaeology of the Pacific generally.

11. Discussion in Finney, *Voyage*, chapter 3.

12. Beaglehole, *Journals*, p. 154n.

13. See Finney, *Voyage*, for extended discussions, upon which I drew here.

14. Terry L. Hunt and Carl P. Lipo, "Late Colonization of Easter Island," *Science*, vol. 311 (2006), pp. 1603–6. See also: Atholl Anderson et al., "Prehistoric Maritime Migration in the Pacific Islands: An Hypothesis of ENSO Forcing," *The Holocene*, vol. 16, no. 1 (2006), p. 1–6.

Chapter 11: The Flying Fish Ocean

1. Louis Klopsch of the *Christian Herald*, 1900. Quoted by Mike Davis, *Late Victorian Holocausts* (New York: Verso, 2001), p. 170. Davis provides a brilliant and devastating analysis of late nineteenth-century tropical famines, which I drew on for this section of the chapter.

2. J. E. Scott, *In Famine Land* (New York: Harper Brothers, 1904), pp. 2–3.

3. Julian Hawthorne, *The Memoirs of Julian Hawthorne, Edited by His Wife, Edith Garrigues Hawthorne* (New York: Macmillan, 1938), p. 295. Reuter's special famine commissioner, Francis Merewether, who was in the famine areas a few months before Hawthorne, shocked the reading public of the day with his graphic descriptions of the victims in his *A Tour Through the Famine Districts of India* (London: A. D. Innes, 1898).

4. Quotes from Julian Hawthorne, "India Starving," *Cosmopolitan*, vol. 23, no. 4 (1897), pp. 379–82.

5. The word "Nilometer" probably originated with early nineteenth-century French scholars on the Napoleonic expedition of 1798.

6. See William Popper, *The Cairo Nilometer: Studies in Ibn Taghri Bardi's Chronicles of Egypt*. Vol 1 (Berkeley: University of California Press, 1951). The Nilometer is well worth a visit, but is off the regular tourist beat. It is easily accessible by taxi from Cairo.

7. Fekri Hassan, "Environmental Perception and Human Responses in History and Prehistory." In Roderick J. McIntosh, Joseph A. Tainter, and Susan Keech McIntosh, eds. *The Way the Wind Blows: Climate, History, and Human Action* (New York: Columbia University Press, 2000), pp. 121–40. Also, more controversially, R. S. Herring, "Hydrology and Chronology: The Rodah Nilometer as an Aid in Dating Interlacustrine History." In J. B. Webster, ed., *Chronology, Migration and Drought in Interlacustrine Africa* (New York: Africana, 1979).

8. D. Verschuren, K. R. Laird, and B. F. Cumming, "Rainfall and Drought in Equatorial East Africa During the Past 11,000 years," *Nature*, vol. 403 (2000), pp. 410–14.

9. Graham Connah, *African Civilizations: An Archaeological Perspective*, 2nd ed. (Cambridge, Eng.: Cambridge University Press, 2001). Chapter 6 offers an authoritative perspective on the East African coast.

10. W. H. Schoff, ed., *The Periplus of the Erythraean Sea: Trade and Travel in the Indian Ocean by a Merchant of the First Century* (New York: Longmans Green, 1912), section 57. Hippalus, a Greek skipper from Alexandria, voyaged from Arabia to India and back in a single season in the first century B.C. He sailed along the Arabian coast, then "by observing the position of the ports and the character of the sea, discovered a route across the ocean." Soon the southwesterly monsoon became known as the Hippalus wind.

11. Revelation 21:10.

12. Discussed at length in Ian Blanchard, *Mining, Metallurgy and Minting in the Middle Ages* (Stuttgart: Franz Steiner, 2001).

13. Ibid., p. 127.

14. Quoted in Ralph Abercromby, *Weather: A Popular Exposition of the Nature of Weather Changes from Day to Day* (London: Kegan Paul, 1887), p. 234. The Subandhu quote that immediately follows can be found in Khushwant Singh, "The Indian Monsoon in Literature." In Jay. S Fein and Pamela L. Stephens, eds., *Monsoons* (New York: John Wiley, 1987), p. 45.

15. For a discussion of the changing landscape and agriculture, see Sumit Guha, *Environment and Ethnicity in India 1200–1991* (Cambridge, Eng.: Cambridge University Press, 1999), chapter 2. Quote from W. H. Sykes, "Special Reports on the Statistics of the Four Collectorates of the Dukhun," *Reports of the Seventh Meeting of the British Association for the Advancement of Science* (1837), p. 226.

16. Quoted from Stanley Wolpert, *A New History of India* (New York: Oxford University Press, 2000), p. 105.

17. Annette S. Beveridge, ed., *Babur Nama* ("Memoirs of Babur") (Delhi: Oriental Books, 1970), p. 488. Babur, who reigned as Mughal emperor from 1526 to 1530, was known as the Master of Hindustan. He was a Timurid prince, who defeated the last of the Delhi Mongol sultans. "Mughal" is the Persian word for Mongol.

18. The general descriptions in this section are based on Michael Coe, *Angkor and the Khmer Civilization* (London: Thames & Hudson, 2003), and Charles Higham, *The Civilization of Angkor* (Berkeley: University of California Press, 2004).

19. This section is based on Roland Fletcher, "Seeing Angkor: New Views on an Old City," *JOSA*, vols. 32–33 (2000–2001), pp. 1–27. Also: Richard Stone, "The End of Angkor," *Science*, vol. 311 (2006), pp. 1364–68.

Chapter 12: China's Sorrow

1. Confucius, *Analects*, book 18. Translation at http://classics.mit.edu/Confucius/analects.html.

2. Loess, from the German word *lö'ss* or *lösch*, meaning "loose," is wind-blown dust of glacial origin, which formed huge deposits in areas far from retreating ice sheets in parts of the central and northwestern United States, in central and eastern Europe, and in northeastern China. For northern Chinese agriculture, see Philip C. C. Huang, *The Peasant Economy and Social Change in North China* (Stanford: Stanford University Press, 1985).

3. Q.-S. Ge, J.-Y. Zheng, and P.-Y. Zhang, "Centennial Changes of Drought/Flood Spatial Pattern for Eastern China over the Last 2,000 Years," *Progress in Natural Science*, vol. 11, no. 4 (2001), pp. 280–87. See also discussion in National Research Council, *Surface Temperature Reconstructions for the Past 2,000 Years* (Washington, D.C.: National Academies Press, 2006). p. 41.

4. Wei-Chyung Wang and Kerang Li, "Precipitation Fluctuation over a Semi-arid Region in Northern China and the Relationship with El Niño/Southern Oscillation," *Journal of Climate*, vol. 3 (1990), pp. 769–83.

5. George Cressey, *China's Geographic Foundations: A Survey of the Land and Its People* (New York: McGraw-Hill), pp. 84–85.

6. Guoqiang Chu et al., "The 'Mediaeval Warm Period' Drought Recorded in Lake Huguangyan, Tropical South China," *The Holocene*, vol. 12, no. 5 (2002), pp. 511–16.

7. Lonnie G. Thompson, "Climatic Changes for the Past 2000 Years Inferred from Ice-Core Evidence in Tropical Ice Cores." In Philip D. Jones, Raymond S. Bradley, and Jean Jouzel, *Climatic Variations and Forcing Mechanisms of the Last 2000 Years* (New York: Springer, 1996), pp. 281–96.

8. Gergana Yancheva et al., "Influence of the Intertropical Convergence Zone on the East Asian Monsoon," *Nature*, vol. 445 (2007), pp. 74–77.

9. Cheng-Bang An et al., "Climate Change and Cultural Response around 4000 cal yr B.P. in the Western Part of Chinese Loess Plateau," *Quaternary Research*, vol. 63, no. 3 (2005), pp. 347–52.

10. Francis H. Nichols, *Through Hidden Shensi* (New York: Scribners, 1902).

11. Quotes in this paragraph from ibid., pp. 229, 231.

12. Quotes in this paragraph from ibid., pp. 242, 245.

13. Scenario based on a generalized description by Morris Rossabi, introduction to Morris Rossabi, ed., *China Among Equals* (Berkeley: University of California Press, 1983), p. 2.

14. On the T'ang, see introduction to Herbert Franke and Denis Twitchett, eds., *The Cambridge History of China*. Vol. 6 (Cambridge, U.K.: Cambridge University Press, 1994), pp. 1–42.

15. On Khitan, see Denis Twitchett and Klaus-Peter Tietze, "The Liao." In Franke and Twitchett, *The Cambridge History*, vol. 6, pp. 43–153.

16. Figures from Herbert Franke, "The Chin Dynasty." In Franke and Twitchett, *The Cambridge History*, vol. 6, p. 292.

Chapter 13: The Silent Elephant

1. Isak Dinesen (Karen Blixen), *Out of Africa* (New York: Random House, 1938), p. 15.

2. D. C. Pederson et al., "Medieval Warming: Little Ice Age, and European Impact on the Environment During the Last Millennium in the Lower Hudson Valley, New York," *Quaternary Research*, vol. 63, no. 2 (2005), pp. 238–49.

3. Al Gore, *An Inconvenient Truth* (Emmaus, Pa.: Rodale Books, 2006).

4. George Orwell, "Shooting an Elephant" (1936); available at http://www.online-literature.com/orwell/887/.

5. http://www.ita.org contains full information, including archives.

6. The Hadley study uses the widely accepted Palmer Drought Severity Index (PDSI) to measure drought levels. "Modelling the Recent Evolution of Global Drought and Projections for the 21st Century with the Hadley Center Climate Model," *Journal of Hydrometeorology*, vol. 7, no. 5 (2006), pp. 1113–25.

7. http://www.unep.org is the home page of the United Nations Environment Program, on whose Web site the latest, and constantly changing, statistics can be found.

8. Mike Davis, *Late Victorian Holocausts* (New York: Verso, 2001), preface.

9. Figures from UNESCO Second World Water Development Report, 2005.

10. Brian Fagan, *The Long Summer* (New York: Basic Books), p. 252.

Index

A Note on the Author

Brian Fagan is emeritus professor of anthropology at the University of California–Santa Barbara. Born in England, he did fieldwork in Africa and has written about North American and world archaeology, and many other topics. His books on the interaction of climate and human society have established him as a leading authority on the subject; he lectures frequently around the world. He is the editor of *The Oxford Companion to Archaeology* and the author of *Fish on Friday: Feasting, Fasting, and the Discovery of the New World*; *The Little Ice Age*; and *The Long Summer*, among many other titles.